Homeowner Manual
A Template for Home Builders

Second Edition

Carol Smith

Home Builder Press®
National Association of Home Builders
1201 15th Street, NW
Washington, DC 20005-2800
(800) 223-2665
www.builderbooks.com

Homeowner Manual: A Template for Home Builders, Second Edition
ISBN 0-86718-517-1

© 2001 by Carol Smith and Home Builder Press®
of the National Association of Home Builders
of the United States of America

All rights reserved. No part of this book may be reproduced or utilized in any form or by any means, electronic or mechanical, including photocopying and recording, or by any information storage and retrieval system in a for-sale publication without permission in writing from the publisher. However, purchasers *do* have permission from the publisher and author to make copies of this publication, or some version of this publication, for distribution to their home-buying customers as part of a customer service program.

Printed in the United States of America

Cataloging-in-Publication Data available from the Library of Congress

Disclaimer
This publication is designed to provide accurate and authoritative information in regard to the subject matter covered. It is sold with the understanding that the publisher is not engaged in rendering legal, accounting, or other professional service. If legal advice or other expert assistance is required, the services of a competent professional person should be sought.
—From a Declaration of Principles jointly adopted by a Committee of the American Bar Association and a Committee of Publishers and Associations.

For further information, please contact:
Home Builder Press®
National Association of Home Builders
1201 15th Street, NW
Washington, DC 20005-2800
(800) 223-2665
Check us out online at: www.builderbooks.com

12/01 B. Minich/Data Reproductions 2000

About the Author

Carol Smith is the leading customer relations expert for home builders. Her 24 years of front-line experience with customers are immediately apparent in her realistic and practical approach. She has performed over 700 buyer orientations, held the posts of superintendent, custom home sales manager, and vice president of customer relations. In 1999 she initiated Customer Relations Professionals (CRP), an international association that provides education and recognition to customer service professionals.

Since 1986 Smith has presented hundreds of educational programs to builders and associates in the United States and abroad, including NAHB's International Builder's Shows, regional conferences, seminars sponsored by the Home Builders Institute (HBI), Custom Builder Symposiums, and Remodelers' Shows. She developed the curriculum for the Home Builders Institute's full-day Customer Service course.

She launched her newsletter, *Home Address* in 1986 and devotes it exclusively to homebuilding service issues. Smith is an award-wining columnist for *Custom Home* magazine and has written dozens of articles for such publications as *Builders Management Journal, Premier Homes, Building Homes and Profits*, and *Builder* magazine.

She has written five books, all published by Home Builder Press, National Association of Home Builders. She expanded three of her books into the two-volume set, Customer Relations Handbook for Builders. She has also written *Building Your Home: An Insider's Guide; Dear Homeowner: A Book of Customer Service Letters*; and a series of 12 customer service brochures for builders. Her most recently completed project involved writing the script for a video, *Building a Home: Dream to Reality*, produced by the Home Builders Association of Greater Kansas City. It was based on her book *Building Your Home: An Insider's Guide.*

She is has been a licensed Colorado real estate brokers since Colorado Broker since 1988.

Acknowledgments

Reviewers

The following people reviewed the outline and/or all or part of the manuscript for *Homeowner Manual: A Template for Home Builders*: James L. Anderson, ACS, Summerville, SC; Michael Crum, Construction Plus of Alabama, Montgomery, AL; Michael C. Golder, Barrington Homes, Apex, NC; Robert Hankin, Prefab City, Inc., Poughkeepsie, NY; O. Pete Kotarides, Kotarides Companies, Inc., Norfolk, VA; Joanna Mahoney, Building News Bookstore, Anaheim, CA; Paul Oliver, Oliver Custom Homes, Austin, TX; Bob Whitten and Laura Carr, Bella Vista, AR; and William Young, NAHB Director of Consumer Affairs, Washington, DC.

Book Production

Homeowner Manual: A Template for Home Builders, second edition is produced under the general direction of Jerry Howard, NAHB Executive Vice President and CEO, in association with NAHB staff members Rob Pflieger, Senior Staff President, Public Affairs; Greg French, Assistant Staff Vice President, Public Affairs; Charlotte McKamy, Publisher, Home Builder Press; Doris M. Tennyson, Senior Acquisitions Editor; David Rhodes, Art and Production Director; and Toral Patel, Assistant Editor; Erica Orloff, project manager; and Maryanne Orloff, proofreader.

Disclaimer

Warning—This publication is designed to present information to builders as an aid to preparation of a homeowner manual. It should be used merely as a starting point and guide for creating a manual that is consistent with each individual builder's contract documents and company policies, procedures, and philosophy. It is not to be regarded as providing opinions or advice for any individual project. Users of this publication should recognize that the maintenance guidelines and the limited warranty language found in this publication are for illustration purposes only; they may vary in a particular case because of local or regional differences in construction codes, techniques and materials, and environmental conditions. Builders should work with their attorneys to prepare contract and warranty documents that meet their particular needs. Builders should also be sure to advise customers to follow manufacturers' instructions for maintenance, where applicable.

How to Use This Model

Customers come to your company with a variety of expectations. Their satisfaction increases when you carefully align those expectations with your procedures, product, and service. Achieving this goal begins with forthright and comprehensive information.

The documents accompanying the purchase of a new home often overwhelm customers: purchase agreement, addenda, no carbon required (NCR) forms, brochures, disclaimers, disclosures, change orders, floor plans, color selections, options list, financial worksheet, mortgage application, closing instructions—the list goes on. Most of this is unread, misunderstood, lost, or ignored by busy customers with no time or patience for fine print.

Yet builders often trace customer dissatisfaction to a lack of buyer understanding of the process and the product. After move-in, warranty issues can cause dissatisfaction. For years builders provided warranty standards and maintenance information at orientation or closing. However, close to moving day buyers are unlikely to read these materials. After move-in, homeowners often send a service request to the builder that typically includes maintenance items. Either the builder provides a grumbling repair or conflict develops when the builder denies service.

In the wake of misunderstandings about change orders, site visits, quality, and delivery dates during the building process, and confusion over maintenance and warranty issues after move-in, builders now recognize their responsibility to guide their customers more effectively. A comprehensive manual delivered at the beginning of the process informs buyers about normal events of the building process and serves as a reference after move-in. This educational approach has been proven to work—increasing homeowner satisfaction dramatically.

This template offers a base from which to develop such a manual for your company. The result can change the way you do business and improve the relationships you have with your buyers. Tailoring and implementing a manual into your system can be accomplished in seven steps.

1. Review and compare.

    - Meet with key personnel to present the concept, provide an overview of the customizing process, and establish a schedule. For most companies, customizing this manual takes 60 to 90 days.
    - Ask your staff and trades to read the manual or at least the sections relevant to them. Their task is to compare the material in the template to current procedures. You are preparing to change the manual to reflect your methods or to change your procedures to match the template, if you prefer. Compile their feedback, section by section.

2. Revise the content using the disk that comes with the template.

   - Begin with the commands to "search and replace" and incorporate your company name where the template says [Builder]. You might insert your company forms in place of the samples included in the template.
   - Delete entries that do not apply to your product—such as the entry about mildew if you build in an arid climate or the entry about termites if they are not a concern in your region.
   - Change procedures as necessary to reflect your methods, such as the steps in the buyer selection process. Adjust terms; for instance, does your region use a deed of trust or a mortgage? Does an engineer design your foundations? Not all regions have this requirement. If the permitting process takes 8 weeks in your area, mention this in your manual. Adjust and emphasize as needed to make this template into your company's homeowner manual.
   - In particular, watch time frames for buyer activities, for example, applying for a loan and making selections, and builder commitments, such as completion date notice and response time on orientation items. Revise care and maintenance entries to reflect the manufacturer's recommendations for items in your homes.
   - Adjust warranty criteria and repairs to meet your particular commitment. For example, is lack of air conditioning an emergency in your area? In some it is not. An excellent resource for performance guidelines is the NAHB publication, *Residential Construction Performance Guidelines for Professional Builders and Remodelers*.
   - Add entries that address specific items relevant to your product, for example, maintenance and warranty guidelines addressing particular plant materials used in builder-installed landscaping. Add energy-saving tips appropriate to your climate and product. Will homeowner association documents be included in this volume or delivered in a second binder? Do you want to add a community information section?

3. Make formatting decisions.

   - Some companies add clip art or digital photos, others add color accents or even cartoons. Keep the image of your company in mind as well as the significance of your product. You want your manual to be interesting and friendly without being too casual or flippant. The style, size, and quality of any clip art you select should be consistent. Resist the temptation to use several different styles simply because they are available.
   - Custom-printed index tabs—done in company colors—can cost less than commercially purchased, numbered tabs. They also look much more professional and make using the manual more convenient.
   - Printing the bullets that summarize each section on the index tabs can serve you well. The sales counselor can easily review the key points by turning to each of the tabs. Printed together in a single handout, they also provide prospective buyers with a concise overview of your processes.

- Consider adding vinyl page protectors, pocket folders, or envelopes for storing additional documents, manufacturer brochures, or buyer-selected color samples.
- Select a good quality paper and a binder to hold your manual—again coordinating colors. A three-ring binder works well as a cover. You can incorporate changes economically to update inventory copies. These binders can be custom printed, although the style with a clear vinyl pocket in the front (available at office supply warehouses for a few dollars each) and a printed insert can work well.
- Consider the style and format you use for page numbering. If you elect to print on both sides of a slightly heavier paper, centered page numbers work well. By numbering each section individually, the impact of updates that alter page numbering is minimized.
- The final step of editing is checking page numbers in the table of contents against the revised pages. Some builders elect to number the entries in the Caring for Your Home section. For example, Air Conditioning is number 1 and each subhead is 1.1, 1.2, 1.3, and so on.
- Print a draft for final review and proofreading. Make any needed corrections, and you are ready for production.

4. Establish your inventory. Depending on your volume, you may have several hundred copies made by a printer. Smaller numbers can be printed individually, inserting the buyers' names throughout the manual.

5. Label this first edition with the date you begin using it and file a master for future reference. As revised editions are completed in the future, repeat this step.

6. Once the manual is complete and ready to use, implement it with enthusiasm:

- Ask that all staff read the final edition.
- Display your manual at the sales site and wherever customers make selections.
- Mention it in the sales presentation. "Mr. and Mrs. Jones, when you purchase one of our homes, you'll receive a copy of our homeowner manual. It contains a lot of useful information about our process and your new home."
- Deliver the buyers' copy at contract.
- Review it briefly, pointing out the overall organization and topics covered.
- Recommend that buyers read through the Caring for Your Home section prior to making selections. Their understanding of maintenance tasks involved with various features and finish materials may influence their choices.
- Ask that buyers acknowledge receipt of the manual, this can be mentioned in the contract or on the buyers' contract checklist.
- Recommend that buyers bring their manual to all meetings with the builder and store documents and even color samples in it for reference. Remind buyers to bring their manual to meetings again when you schedule those meetings.
- Create high visibility. Ask all staff to bring it to meetings with buyers, use it themselves, and refer to it in conversations with buyers. Remind buyers to review the upcoming section at the end of routine meetings.

- During random visits, remind buyers to read the manual. Ask if they have questions about it, and refer to it when you answer their questions.
- Discuss the manual in more detail at orientation, focusing on how to locate maintenance and warranty information.
- Carry a copy to warranty inspections as a reference (tempered with common sense, of course). In extreme cases, quote it in follow-up letters to homeowners.

7. Ask each staff member to start a "Manual Revisions" file and make note of their ideas for future updates. Suggestions from homeowners can also be incorporated in annual updates.

8. *Note to custom builders:* Special sections that were created to meet the unique needs of custom builders are included in Appendix A. These sections can be incorporated into your custom homeowner manual. Refer to Appendix A for: Written Agreements, Budget and Financing, New Home Design and Selections, and Construction of Your Home.

9. Appendix B offers an optional approach to Warranty Service. This section can easily be incorporated into your homeowner manual.

You are now ready to use the manual with customers.

What about buyers who are already under contract or homeowners who have already closed and moved into their homes? Depending on the numbers involved, builders have delivered copies to homeowners who have already gone to closing and those under contract. Some builders stock the sales office with extra copies and provide them to veteran homeowners who ask for them. Present your new manual not as changing anything already agreed to, but simply as a more complete and better-organized presentation. If any conflicts exist between what you say in your new manual and the original signed agreement you have with a buyer, the original signed agreement applies, unless you both agree to follow the new version.

What about buyers who don't read the manual? If your company integrates the manual through the process, most buyers will accept that the manual is the authority for daily decisions at your company. Most of us have not read the dictionary cover to cover, yet we accept the authority of it when we need to look up how to spell a word. This is your goal with home buyers. Even buyers who do not read the manual are more likely to conclude they have been treated fairly when you refer them to it later. The trust that grows from forthright communication is vital to long-term homeowner satisfaction.

[Builder]

Homeowner Manual

[Builder]

Homeowner Manual

Receipt

Congratulations on your decision to build a new home!

[Builder] is proud to deliver this copy of our homeowner manual to you as part of the purchase agreement materials for your new home:

Date _____

Community _____

Floor plan _____

Address _____

Legal _____

Please acknowledge for our records that you received this manual:

Purchaser_____   Date_____

Purchaser_____   Date_____

[Builder] Homeowner Manual

___

Congratulations on your decision to purchase a new home from [Builder]. We share your excitement about your new residence and look forward to having you work with us to have your home built.

[Builder] designed this *Homeowner Manual* to assist you during and after the purchase of your home. The information presented here will answer many questions and prepare you for each step of the new home experience, making this exciting time easier.

In addition to guiding you through the process of purchasing and building, this manual provides you with maintenance guidelines and a description of our limited warranty program, component by component.

Please take time to review this material thoroughly. Note the amount of detail we have provided. Your new home will receive the same attention to detail.

Please bring this manual to all scheduled meetings. As we progress, you will add items to it. When complete, your manual will provide a useful record of information about your new home.

If you need clarification or additional details about any topic discussed, please give us a call. We are delighted to welcome you as part of the [Builder] family and are always ready to serve you.

Sincerely,

[Builder]

Contents

1	**Introduction**	9
	[Builder]	10
	What Happens Next?	11
	Who's Who?	13
2	**Purchasing Your Home**	15
	Purchaser Checklist	16
	Purchase Agreement	16
	Addenda	16
	Purchaser Checklist	18
3	**Arranging Your Loan**	21
	Loan Application Checklist	22
	Loan Application Paperwork	24
	Loan Underwriting	25
	Loan Lock	26
	Loan Closing	26
	Down Payment Worksheet	27
	Monthly Payment Worksheet	28
4	**New Home Selections**	29
	Standard Features	30
	Optional Features	30
	Custom Features	30
	Selection Hints	31
	Selection Locations	33
	Buyer Start Order	34
	Buyer Start Order	35
	Change Orders	36
	Change Order	38
5	**Construction of Your Home**	41
	Preconstruction Conference	42
	Start of Construction	42
	Preconstruction Agenda	43
	Safety	44
	Frame Tour	44

[Builder] Homeowner Manual

	Locks and Keys	45
	Plans and Specifications	45
	Quality	47
	Private Home Inspectors	48
	Single Source	48
	Trade Contractors	49
	Schedules	49
	Construction Sequence	50
	Our Customer Wants to Know	53
6	**Homeowner Orientation**	**55**
	Scheduling	56
	Last-Minute Activity	56
	Preparation	56
	Completion of Items	58
	Feedback on Orientation Items	59
	Orientation Forms	60
7	**Closing on Your Home**	**63**
	Date	64
	Location	64
	Documents	64
	"The Final Number"	66
	Preparation	66
	House Keys	67
	Garage Door Opener Operators	67
	Mailbox Keys	67
	First Mortgage Payment	67
	Storing Documents	67
	Utility and Community Services	68
	Moving Hints	70
8	**Caring for Your Home**	**73**
	Homeowner Use and Maintenance Guidelines	74
	[Builder] Limited Warranty Guidelines	75
	Warranty Reporting Procedures	76
	Warranting Item Processing Procedures	79
	Help Us to Serve You	79
	Warranty Service Summary	83
	Warranty Hours	83
	Appliances	83
	Emergency	83
	Nonemergency	83

Storm Damage or Other Natural Disaster 83
Fire Prevention 84
Train Family Members 84
Practice Prevention 84
Extended Absences 86
Plan in Advance 86
As You Leave 86
Energy and Water Conservation 88
Heating and Cooling 88
Water and Water Heater 89
Appliances 89
Electrical 90
Maintenance 90
Appliance Service 91
Home-Care Supplies 92
Maintenance Schedule 93
Air Conditioning 94
Alarm System 96
Appliances 97
Asphalt 97
Attic Access 98
Brass Fixtures 99
Brick 100
Cabinets 100
Carpet 102
Caulking 105
Ceramic Tile 106
Concrete Flatwork 107
Condensation 110
Countertops 111
Crawl Space 112
Dampproofing 113
Decks 113
Doors and Locks 115
Drywall 117
Easements 118
Electrical Systems 119
Evaporative Cooler 122
Expansion and Contraction 123
Fencing 123
Fireplace 125
Foundation 127
Garage Overhead Door 128
Gas Shut-Offs 130
Ghosting 130
Grading and Drainage 131

[Builder] Homeowner Manual

Gutters and Downspouts 133
Hardware 134
Hardwood Floors 135
Heating System: Gas Forced Air 137
Heating System: Heat Pump 142
Humidifier 144
Insulation 144
Landscaping 145
Mildew 150
Mirrors 150
Paint and Stain 150
Pests and Wildlife 152
Phone Jacks 153
Plumbing 153
Property Boundaries 160
Railings 161
Resilient Flooring 161
Roof 163
Rough Carpentry 165
Shower Doors or Tub Enclosures 167
Siding 167
Smoke Detectors 168
Stairs 169
Stucco 169
Sump Pump 170
Swimming Pools 171
Termites 172
Ventilation 173
Water Heater: Electric 174
Water Heater: Gas 175
Windows, Screens, and Sliding Glass Doors 177
Wood Trim 179
Warranty Service Request 181
One-Time Repairs 182
Homeowner Comments 183

[Builder] Homeowner Manual

Section 1: Introduction

✔ [Builder]—some background on our company

✔ What Happens Next?—an overview of the major steps in the home buying process

✔ Who's Who?—names and contact information for key people who will assist you in this process

Introduction

[Builder]

Develop a one-page presentation of background information about your company to insert here. Include such details as when the organization was founded, its philosophy, awards and accomplishments, or unusual traits for which it is known. Perhaps your company has won awards for innovative designs. Maybe you are well-known for energy-conserving features and "green build" construction techniques.

Consider mentioning community involvement or contributions. Have your employees worked on homes with Habitat for Humanity? Do you sponsor the local elementary school baseball team?

Material from brochures or other marketing materials may contribute items to include. Some companies include their mission statement, or use this space to introduce key players in the organization. A collage of photos of an established team makes the people of the organization real to the buyers and may be just right for your company.

What Happens Next?

An Overview of Your New Home Experience

Purchasing a new home is an exciting experience. The process is also complex, with many details to be decided and arranged. While [Builder] is building your new home, you participate by taking care of several important aspects of your purchase.

Building a new home is an investment of your money, your emotions, and your time. Many of the tasks will require your attention during regular business hours, Monday through Friday, usually between 8:00 a.m. and 5:00 p.m.

The chronological list that follows outlines the events that typically take place in the purchase of a new home and provides an overview of the events that will require your time and attention. Where time frames are specified, you need to observe them in order for us to deliver your home on schedule.

Purchasing Your Home

The purchase agreement and various addenda constitute the legal understanding regarding the purchase of your new home. Please read the purchase agreement and all attachments carefully. As with any legal agreement, you may wish to have your attorney review them. Once all the paperwork is signed, we suggest you insert those documents in Section 2 of this manual, Purchasing Your Home.

Arranging for Your Loan

Once you have signed the purchase agreement, finalizing the details for financing is next. To assist you, we may suggest lenders appropriate for your specific financial situation. Section 3, Arranging for Your Loan, contains hints and information on the loan process.

New Home Selections

New Home Selections, Section 4 of this manual, will assist you in the exciting process of personalizing your new home with your selections.

Construction of Your Home

Several tasks need to be completed prior to the start of construction. Some of these are our job; some are yours. They are described in Section 5, Construction of Your Home. Near the beginning of construction, we will offer to meet with you at a Preconstruction Conference to review plans and specifications one final time. Next we will invite you to tour your new home with us when your home reaches the mechanical stage, just before insulation is installed. Please bring this manual to both of these meetings.

We also expect and welcome your casual visits to the site. Please read Section 5, Construction of Your Home, for guidelines on safety, security, and work in progress.

Homeowner Orientation

The homeowner orientation has two purposes. The first is to demonstrate the features of your home and discuss maintenance and our limited warranty program. Equally important, we want to confirm that we have delivered your new home at the quality level described in our documents and shown in our model homes and with all your selections correctly installed. For detailed information, please review Section 6, Homeowner Orientation.

Closing on Your Home

Closing on Your Home, Section 7 of this manual, describes the documents you will sign and other important details about the closing process. We have included guidelines to assist you in preparing for closing and move-in.

Caring for Your Home

Many of your responsibilities as an owner and [Builder]'s responsibilities under the terms of our limited warranty are discussed in Caring for Your Home, Section 8. Begin now to become familiar with the home maintenance you should provide and our warranty service commitment to you. [Builder] plans two standard contacts with you during the warranty period. These visits and procedures for service outside these standard contacts are described in Section 8.

Your Feedback and Suggestions

Our desire to maintain open communication with you extends through the buying process and after your move-in. In an effort to improve the product and service we provide, we welcome your comments on how we've performed. We survey our customers after move-in. Our goal is to build the best home and the best customer relationship possible. Your feedback helps us reach that goal.

As time passes, if your housing needs change, we are ready at any time to build you another home. We also appreciate your referrals. Our office is always happy to provide you with information about where we are currently building and the products we offer.

Who's Who?

Some Names You Should Know

Two-way communication is vital to a mutually satisfactory relationship. Understanding what is happening and knowing who to contact can smooth the home-buying process. We believe that it is our responsibility to establish and maintain clear lines of communication. The professionals listed below are glad to assist you or find the answers to your questions. A plastic sheet follows as a convenient location for business cards, as well.

Builder
[Builder]
<Address>
<Phone>
<Fax>
<E-mail>
<Hours>

Sales Counselor
<Name>
<Address>
<Phone>
<Fax>
<E-mail>
<Hours>

Superintendent
<Name>
<Address>
<Phone>
<Fax>
<E-mail>
<Hours>

Lender
<Name>
<Address>
<Phone>
<Fax>
<E-mail>
<Hours>

Interior Designer
<Name>
<Address>
<Phone>
<Fax>
<E-mail>
<Hours>

Title Company
<Name>
<Address>
<Phone>
<Fax>
<E-mail>
<Hours>

Warranty Office
<Name>
<Address>
<Phone>
<Fax>
<E-mail>
<Hours>

Real Estate Agent
<Name>
<Address>
<Phone>
<Fax>
<E-mail>
<Hours>

Note to Builder:

Insert a plastic page for business cards here.

Section 2: Purchasing Your Home

✔ Purchaser Checklist–your opportunity to confirm we have communicated clearly and have delivered all necessary documents

✔ Purchase Agreement–a brief description of each of the documents you will receive

✔ *Purchaser Checklist*–a copy of the form you sign at the end of your purchase agreement session

Purchasing Your Home

You will use several standard forms when you buy your new home. These include the purchase agreement and several addenda. The purchase agreement becomes binding only when all parties have signed all forms and attachments.

If you are new to the United States, [Builder] welcomes you and understands that you may be unfamiliar with our business procedures and traditions. We will gladly discuss any questions you may have about the U.S. business practices we will be following.

Purchaser Checklist

This sheet confirms that we delivered all necessary documents and discussed key topics in order to prevent surprises. Our experience shows that the new home process progresses more smoothly with good communication. To be certain that we have been clear in explaining our purchase agreement and that we have called your attention to clauses or topics that have caused confusion in the past, we will ask you to sign this confirmation at the end of the meeting.

Purchase Agreement

The purchase agreement is the legal document that represents your decision to purchase a home. It describes your home (both a legal description and the street address), financing information, homeowner association information, if applicable, and additional legal provisions. We recommend that you read these documents carefully. In particular, please take note of the topics listed on our Buyer's Checklist which we will discuss with you prior to your signing your purchase agreement.

Several exhibits are typically attached to the purchase agreement. The features of the community determine the specific items, but the list below is typical.

Addenda

Exhibit A

Materials and Specifications list materials and methods to be used in construction of your home.

Exhibit B

Allowance Schedule lists categories and amounts included in the price of your home for finish materials you select.

Exhibit C

Selection Sheets outline details of your finish material choices, such as color, brand, model, and so on. Please plan to complete these within 30 days of signing your contact. See Section 4, New Home Selections, for more information.

Exhibit D

[Builder] Limited Warranty, a specimen copy for your study, with the actual warranty executed at closing.

Exhibit E

Homeowner Association Documents, where applicable.

Homeowner Manual

This book is your Homeowner Manual. It will guide you through the building process and serve as a useful reference after your move in.

Community

Our community information materials contain specific documents and disclosures about the local community.

Purchaser Checklist

Purchasers _____ Date_____

Your signature below confirms that we have delivered the following items to you:

___ Purchase agreement
___ Purchase agreement addenda
___ _____
___ _____
___ Materials and specifications for your floor plan
___ Allowance schedule
___ Selection sheets for your floor plan
___ Limited warranty
___ Homeowner association documents, if applicable
___ Homeowner manual
___ Receipt for your deposit, $_____

That we discussed the following clauses from your purchase agreement:

- Allowances
- Reimbursable expenses
- Financing
- Commence and complete construction
- Change orders: procedure and schedule
- Conformance with plans and specifications
- Plan ownership
- Site visits: procedures and safety
- Noninterference
- Inspection and acceptance: orientation
- Site clean-up
- Insulation notice
- Radon disclosure
- Limited warranty: written lists for non-emergency items; standard checkpoints at 60 days and 11 months; emergency items by phone
- Homeowner Association
- Settlement or closing: target delivery date and delivery date updates
- Possession
- Insurance
- Default or termination
- Alternative dispute resolution
- Co-op broker
- Entire agreement

And that we discussed the following topics to expedite communications during the process:

   ___   Scheduled construction meetings: Preconstruction conference, pre-drywall tour
   ___   Buyers' preferred contact:
           Monday – Friday _____ Phone_____
           Saturday _____ Phone_____
   ___   [Builder]'s preferred contact:
           Monday through Friday, 7 a.m. to 6 p.m. at <phone>
           Saturday, 9 a.m. to 1 p.m. at <phone>

Other

_____
_____
_____
_____
_____
_____

Purchaser _____ Date_____

Purchaser _____ Date_____

Builder _____ Date_____

Note to Purchaser:

Insert the completed purchase agreement and addenda here.

Section 3: Arranging Your Loan

✔ Loan Application Checklist–lists the documents and information typically needed to complete the loan application form

✔ Loan Application Paperwork–an overview of the forms involved in processing your application

✔ Loan Underwriting–key points to be aware of regarding the loan approval process; take special note of contingencies that may apply

✔ Loan Lock–lock your loan only after [Builder] has provided you with a written delivery date confirmation

✔ Loan Closing–avoid changes to your financial circumstances to protect your loan approval

✔ *Down Payment Worksheet*–to assist you in determining the amount you have available for your down payment

✔ *Monthly Payment Worksheet*–to assist you in estimating the monthly payment amount for your new home mortgage

Arranging Your Loan

The first items you'll need to take care of are selecting a lender and completing a mortgage application. Plan to accomplish this within 5 business days of signing your purchase agreement. Take the completed purchase agreement with you when you first visit your lender.

Your lender's job is to understand your particular financial circumstances completely. You will review all information on the application at your meeting with the loan officer. A situation rarely arises that your loan officer has not encountered in the past. Do not hesitate to discuss any questions you have regarding your assets, income, or credit. By providing complete information, you prevent delays or extra trips to deliver documents.

Loan Application Checklist

The amount of documentation and information required for a mortgage can seem overwhelming. You can facilitate the application process by collecting as much of the needed information as you can before your appointment.

The checklist that follows is a general guide to assist you with the loan application. Some of the items listed may not apply to you, and your lender will probably request some items that we have not mentioned, but this list will get you off to a good start.

Credit Report

Please note that you will be asked to pay for a credit report and an appraisal upon signing the application.

Property Information

__ The purchase agreement will include the legal description of the property and the price.

Personal Information

__ Social Security number and driver's license for each borrower
__ Home addresses for the last two years
__ Divorce decree and separation agreements, if applicable
__ Trust agreement, if applicable

Income

__ Most recent pay stubs
__ Documentation on any supplemental income such as bonuses or commissions
__ Names, addresses, and phone numbers of all employers for last two years
__ W-2s for last two years
__ If you are self-employed or earn income from commissioned sales, copies of last two years of tax returns with all schedules and year-to-date profit and loss for current year, signed by an accountant
__ Documentation of alimony or child support, if this income is considered for the loan

Real Estate Owned

__ Names, addresses, phone numbers, and account numbers of all mortgage lenders for the last seven years
__ Copies of leases and two years of tax returns for any rental property
__ Market value estimate

Liquid Assets

__ Complete names, addresses, phone numbers, and account numbers for all bank, credit union, 401K, and investment account
__ Copies of the last three month's statements for all bank accounts
__ Copies of any notes receivable
__ Value of other assets such as auto, households goods, and collectibles
__ Cash value of life insurance policies
__ Vested interest in retirement funds or IRAs

Liabilities

__ Names, account numbers, balances, and current monthly payment amounts for all revolving charge cards
__ Names, addresses, phone numbers, and account numbers for all installment debt and approximate balances and monthly payments for such items as mortgages, home equity loans, and auto loans
__ Alimony or child support payments
__ Names, addresses, phone numbers, and account numbers of accounts recently paid off, if used to establish credit

Loan Application Paperwork

Once you have given all preliminary information to your loan officer, your lender sends verification forms to your employers, banks, and current mortgage company or landlord, and also orders the credit report and appraisal. You sign a release to authorize these steps. Your lender will provide you with a Good Faith Estimate and a Truth-in-Lending Disclosure.

Good Faith Estimate

The Good Faith Estimate lists the estimated costs you will incur at closing. Some of the numbers listed on this form are prorations, subject to change based on the actual date of the closing. Others are set fees that should remain the same.

Truth-in-Lending Disclosure

The Truth-in-Lending Disclosure shows the total cost to you, over the term of the loan, for your specific financing. The calculation is based on the assumption that you own the home and make regular payments throughout the term of the loan.

Verification of Employment

The lender sends Verification of Employment (VOE) forms to all employers for the last two years. The employers complete, sign, and return the forms to the lender. The forms show the dates of employment, the amount of money you earned last year, and how much you have earned so far this year. The VOE documents bonuses and overtime you earned.

Verification of Deposit

Verification of Deposit (VOD) forms go to each banking institution listed on your application. The institutions indicate the date you opened each account, average balances for the last three months, and the amount of money you have in each account on the day they complete the form. Any loans or overdraft accounts you have with the bank will also be shown.

Verification of Mortgage

Mortgage companies and landlords complete Verification of Mortgage (VOM) forms. These show the lender how much you owe, the amount of your monthly payment, and whether you make your payments by the due date.

Credit Report

Your credit report shows the amounts of money you owe to each of your creditors, minimum monthly payments, and your payment history. The appraisal confirms the value of the home you are purchasing for you and your lender.

Loan Underwriting

Typically, several weeks pass as these reports and forms are returned to the lender. If any delays are encountered, the loan officer may contact you for assistance. The credit reporting agency may call you to verify that the information they have gathered is correct.

Once the loan processor has collected this standard documentation, you may be asked to write letters describing your assets, income, or credit. Few loans are finalized without requests for additional information just before the package is submitted to the underwriter for final approval. At this point you may become frustrated with the loan process.

Please remember that your lender requests these letters to assist you in obtaining your financing. Do not hesitate to discuss your concerns with your loan officer. Perhaps he or she can provide some additional insight on what may seem to be redundant requests.

Loan Amount Requested

Before the processor submits your file to the underwriters for final approval, he or she will verify the final sales price. Make sure that copies of all addenda such as change orders signed after the original purchase agreement was completed have been sent to the lender. This assists the lender in determining the exact loan amount. If change orders affect the total price after this point, you may have to resubmit your loan application for the higher amount or the lender may ask you to pay for the additional items in cash.

Loan Approval

During your first meeting, you and your lender determine the timing to obtain prequalification. This allows us to start the home even though final approval is still pending. You will discuss additional items that you may need to obtain final loan approval. Several weeks after your first meeting with the lender, you should receive loan approval. If any of the documents requested have not been returned to the lender in a timely manner, approval may take longer.

Contingencies

Loan approvals often carry conditions of approval. The sale of a previous home or proof of funds are two examples. Discuss any concerns you may have about such conditions with your loan officer and obtain any requested documentation as soon as possible. Once all contingencies are met, the final loan can be approved.

Loan Amount Approved

If you qualify for an amount that is less than you requested, ask your loan office what changes might qualify you for a larger loan. Or, consider omitting some items now (a deck or finished basement) and adding them to your home later. Another possibility is to talk to another lender with different programs and different requirements.

Loan Declined

If, after your best efforts, you are not approved for a loan within 45 days of signing your purchase agreement, in accordance with your Purchase Agreement, [Builder] will refund your initial deposit upon your signing a release letter and returning this Homeowners Manual to the sales office.

Loan Lock

The only thing anyone knows for certain about interest rates is that they will change. Do not rely on anyone's predictions regarding rates. Locking your rate prematurely can result in extra expense if your new home is not complete in time to close within the lock period. We are happy to update you throughout the process of construction on the target delivery date. ***Until we reach a point in construction where factors outside our control can no longer affect the delivery date, the decision to lock your loan is at best a gamble.***

Loan Closing

Between the time your loan is approved and the date of your closing, remember that any significant changes in your financial circumstances could impact your loan approval. If your closing occurs more than 30 days after the lender issues your loan approval, the lender may order an additional credit report just prior to the closing date. Changes in your financial circumstances, for example, purchasing a new car or increases in your charge card will appear as a new liability on your updated credit report. Such changes may cause your lender to reconsider your approval. Holding off on such purchases until after closing is best.

Down Payment Worksheet

Available Funds

    Equity in present home $ _____

    Savings, savings certificates   _____

    Investments   _____

    Insurance (cash value)   _____

    Other funds (such as a cash gift)   _____

    Total available funds   _____

    Minus amount you want to keep in savings   _____

    Adjusted Total Available Funds $ _____

Expected Expenses

    Settlement costs (estimate 5 percent of loan) $ _____

    Moving costs   _____

    Landscaping   _____

    Other expected expenses   _____

    Total Expected Expenses $ _____

Down Payment

    Adjusted total available funds $ _____

    Minus total expected expenses   _____

    Amount Available for Down Payment $ _____

Monthly Payment Worksheet

Loan Payment

      Principal and interest $ _____
      Property tax   _____
      Hazard insurance   _____

      Total Loan Payment $ _____

Homeowner Association Monthly Dues $ _____

Estimated Utilities

      Electric $ _____
      Gas   _____
      Water   _____
      Sewer   _____
      Trash collection   _____
      Cable TV   _____
      Security system monitoring   _____

      Total Estimated Utilities $ _____

Monthly Payment

      Loan payment $ _____
      Homeowner association dues   _____
      Estimated utilities   _____

      Total Monthly Payment $ _____

[Builder] Homeowner Manual

Section 4: New Home Selections

✔   Standard Features–confirm your understanding of which features are included in your new home

✔   Optional Features–you can select from many popular options to personalize your new home

✔   Customer Features–[Builder] will consider your requests for custom features with a design/pricing deposit of $200

✔   Selection Hints–reminders to guide you through the selection process

✔   Selection Locations–names and locations of showrooms where you can view selections and options

✔   *Buyer Start Order*–a form you sign confirming all selections and changes are complete and telling us to start your home

✔   Change Orders–[Builder] will consider requests for changes after you sign the Buyer Start Order in accordance with the schedule and fees described here

✔   *Change Order*–a copy of the form that documents any changes, requiring the signatures of all parties and full payment prior to implementation

New Home Selections

Part of the fun of buying a new home is selecting features, finish materials, and colors. You will make some of these choices at the [Builder]'s office and others at our suppliers' showrooms. Location and contact information is included in this sections. As you make choices for your new home, consider your present and future lifestyle. Take into account your family's daily activities, hobbies, and work; the kind of entertaining you do, and your family's holiday traditions.

Standard Features

Each floor plan includes a substantial number of standard features as listed on the standard features sheet available from our sales center. Please review this information carefully to prevent any misunderstandings about which features are included in the base price of your new home. To delete a standard feature, complete and submit a change order with your selection sheets. If you have any questions, your sales counselor will be able to assist you.

Optional Features

Based on feedback from our customers, [Builder] has developed a list of the most popular options that are available for the home plans in your new community. This list and the current pricing of these items is available from your sales counselor. This list is updated regularly based on feedback from our customers and fluctuations in costs.

Our options list is organized by components: cabinet, electrical, floor covering, plumbing, and so on. To include an optional item in your new home, simply list the item on your selection forms.

Custom Features

The possibilities for your new home far exceed the popular ideas we suggest on our options list. In addition to the available options, you may have custom features you want us to consider incorporating into your new home.

Think, dream, imagine, look—we will assist you in any way that we can to make these decisions as early as possible. Please keep in mind that your new neighbors have this same opportunity and may request still other features. We make no claim that we mention or offer every possible idea.

All requests for custom features require a custom design/pricing deposit of $200. The full amount becomes a credit against the cost of the change if you approve the change order. If you decide not to proceed, [Builder] retains the design/pricing deposit.

Selection Hints

[Builder] provides you with selection sheets that list the choices you need to make. Schedule time to visit both our office and our suppliers' showrooms to make your selections as soon as possible. Plan to finalize your selections within 30 days of signing your purchase agreement. Your prompt completion of these selections helps prevent delays caused by backorders.

Informed Choices

We recommend that you review the maintenance tasks and warranty guidelines in Section 8 of this manual prior to making your selection decisions.

Be Thorough

Our selection sheets are very detailed. Fill in all blanks completely. Costly errors arise from assumptions and incomplete selection sheets. After completing this form, double-check all color numbers and names and sign and date each page.

Allowances

Decorating choices that exceed the specified allowances, such as those for floor coverings, countertops, or light fixtures, will require additional payment. Although such amounts can be credited to you at closing and subsequently added to your mortgage, they are not refundable.

Colors

You are welcome to bring cushions or swatches to showrooms to coordinate colors. View color samples in both natural and artificial light to get an accurate impression of the color. Variations between samples and actual material installed can occur. This is because of the manufacturer's coloring process (dye lots) and the fact that over time, sunlight and other environmental factors affect the samples. Some colors will appear different when seen in a large area as opposed to the sample.

Exterior Choices

Your homeowner association and selections your future neighbors have made may limit your choices for exterior finish materials or colors. The sooner you make your selections, the more choices you have. Viewing existing homes is one way to select exterior colors. Selections often look different on a full-size home. Some colors require extra coverage which can impact the cost.

Selection Hold

We reserve the right to place a hold on your selections until your lender has approved your loan and all contingencies are released. If suppliers have discontinued any of your selections, we will contact you and ask you to make an alternate selection within 5 days. Occasionally, a home is already under construction and [Builder] has made some or all of these choices.

Availability

If a selection you make turns out to be unavailable, we will contact you and request that you make a different selection within 5 business days. Because so many choices are offered, [Builder] is unable to predict when a particular manufacturer or supplier may discontinue any particular item. We regret any inconvenience this causes. Similarly, materials readily available when your home is built may not be available in years to come if replacements are needed.

Record of Selections

Please retain your selection sheets for future reference. They are useful for matching paint colors, tile grout, and replacement items in your home.

Selection Locations

Item(s) _____
Contact _____
Company _____
Address _____

Phone _____
Fax _____
Hours _____

Item(s) _____
Contact _____
Company _____
Address _____

Phone _____
Fax _____
Hours _____

Item(s) _____
Contact _____
Company _____
Address _____

Phone _____
Fax _____
Hours _____

Item(s) _____
Contact _____
Company _____
Address _____

Phone _____
Fax _____
Hours _____

[Builder] Homeowner Manual

Buyer Start Order

When you have completed selections for your new home, sign the Buyer Start Order (a copy is included on the next page) which notifies us that we can finalize the orders for your home and schedule the start of construction.

Depending on permitting and trade contractor workload, construction of your new home will begin 2 to 6 weeks after we receive your signed Buyer Start Order. Once you sign the Buyer Start Order, [Builder] orders materials and schedules labor to build your home. Administrative fees apply to any requested changes to your plans and specifications after this point.

Buyer Start Order

Date _____

To: [Builder]

From _____ (Purchasers)

Re _____ (Address of new home)

We have completed our selections and made final decisions on all change requests.

You are hereby authorized to start our home.

We understand and acknowledge that we may request further changes during construction of the home in accordance with the change order schedule for our community.

Each change will include a $200 administrative fee in addition to the cost of the change requested. The custom feature design/pricing deposit will also apply.

Further, should any future change add days to the schedule, payment of construction loan interest for those days will be our responsibility as well and will be added to the cost of the change order.

Purchaser _____ Date_____

Purchaser _____ Date_____

Change Orders

[Builder] uses a change order form (see sample at the end of this section) to describe and document all changes you may request to your new home's plans and specifications. Change orders fall into three categories. You may decide to:

- Add or delete items from the options list after signing your selection sheets
- Change a selection previously ordered
- Personalize your home plans still further with a custom feature

In order to deliver your home as close as possible to the target date, we order many items well in advance of installation. Once a particular item is ordered, making further changes may involve adjusting the planned delivery date and additional costs. By requesting all changes prior to signing the Buyer Start Order, you avoid both.

Processing

When you request a change, the sales counselor will document the request and submit it for approval and, in the case of custom changes, pricing. Pricing of custom change requests typically takes 5 to 10 business days.

Sometimes a seemingly minor change impacts other elements of the home and therefore may come with hidden costs–for example, if you order a ceiling fan, the framing that will hold it is reinforced. If you add a window, framing, drywall, interior and exterior trim, and paint costs may all be affected.

Changes of any kind requested after the cutoff dates for your community include an administrative fee. This is necessary because previously issued paperwork must be canceled and reissued. Errors in this process are a [Builder] responsibility. If the change you request impacts the construction schedule, our pricing will include construction loan interest for the additional days. The cost of deleted items will be credited to you although administrative fees are non-refundable.

Information on pricing and any schedule adjustment is returned to your sales counselor who will then contact you for a final decision. If you elect to proceed with the change, we ask that you sign the change order and make full payment. Change orders that remain unsigned or unpaid become null and void upon the expiration date shown on the change order.

[Builder] Homeowner Manual

For the protection of all concerned, all changes are documented and incorporated into your new home only after

- [Builder] has approved and signed the change
- You have approved, signed, and paid for the change prior to its expiration date
- The applicable building department has approved the change, when applicable

Our contracts with our trade contractors prohibit them from making any changes to plans or specifications without written change order authorization from [Builder].

Cutoff Points for Changes

[Builder] follows a schedule of cutoffs for changes as shown below. [Builder] reserves the right to deny changes you request after these cutoffs.

Changes affecting	Should be made prior to the start of
Foundation	Engineering and permit application
Windows, doors, elevation	Foundation
Mechanical systems, cabinets, appliances	Framing
Texture, hardware, lighting	Mechanical rough-ins
Interior trim and floor coverings	Insulation
Landscape design or materials	Interior trim

[Builder]

Change Order # _____

Purchasers _____   Date_____
Contract dated _____   Plan_____
Address _____   Lot #_____

Description of Change

_____
_____
_____
_____
_____
_____
_____
_____

Design/pricing deposit_____   Expiration date_____
Administrative fee   _____
Cost of change   _____   Delivery date adjustment_____ days
Credit (deleted items) _____
Total   _____

Purchasers have requested the change described above, its costs, and the corresponding adjustment in the construction schedule. By signing this change order, Purchasers agree to pay for this change and acknowledge that the estimated delivery date for the home is revised accordingly. [Builder] will incorporate the change into the home only when the change order has been approved and signed by [Builder], and paid in full by Purchasers. [Builder] has the option of revising the cost, delivery date adjustment, or both if Purchasers have not signed this change order by the expiration date above.

Approved, _____   Purchaser_____
          [Builder]
                                                          Purchaser _____

Date _____   Date_____

Date payment received _____

Note to Home Buyer:

Insert your records of your new home selections here.

Section 5: Construction of Your Home

- ✔ Preconstruction Conference–a meeting to review your plans, selection, changes, and the protocols of the construction process

- ✔ Start of Construction–once you sign the Buyer Start Order, [Builder] attends to several tasks before starting construction

- ✔ Safety–please respect the potentially dangerous nature of a construction site and follow our site visit policies

- ✔ Frame Tour–your second meeting with your builder provides an opportunity to see the quality inside the walls of your new home and confirm that selections and change orders are correct so far

- ✔ Locks and Keys–once you use your house keys, only your keys will open your home

- ✔ Plans and Specifications–no two homes are alike

- ✔ Quality–we monitor work on your home to note and correct any errors that occur and ensure that the home we deliver meets the standards we promised you

- ✔ Single Source–[Builder] selects all personnel and orders all materials that go into your home

- ✔ Trade Contractors–trades people have no authority to make changes without [Builder]'s written change order and are unaware of all the elements in your home; any questions you have should be communicated through your salesperson

- ✔ Schedule–delivery dates are a target until we confirm a closing date in writing; we promise a minimum of 30 days notice

- ✔ Construction Sequence–an overview of the major steps typically followed in building a home

- ✔ *Our Customer Wants to Know*–forms for your convenience, please document any questions you have about your home during construction and forward them to your sales counselor

Construction of Your Home

The construction of a new home differs from other manufacturing processes in several ways. By keeping these differences in mind, you can enjoy observing the construction process as we build your new home.

- As a consumer, you rarely have the opportunity to watch as the products you purchase are created. Your new home is created in front of you.
- You have more opportunity for input into the design and finish details of a new home than for most other products. Our success in personalizing your home depends on effective and timely communication of your choices.
- Because of the time required for construction, you have many opportunities to view your home as it is built, ask questions, and discuss details.

Preconstruction Conference

You will have the opportunity to meet with your builder twice during this process. The first of these is a preconstruction conference (the second is the frame tour described later in this section). Your sales counselor schedules this appointment once all of your selections are completed. This meeting takes approximately 60 to 90 minutes.

The purpose of the preconstruction conference is to conduct a comprehensive review of your final plans and specifications as well as the building process itself. We will discuss such things as site visits, questions, trade contractor communication, change orders, and target delivery date. A copy of our agenda is included on the next page. Please bring any questions you have and this manual with you to this meeting.

Start of Construction

Before construction of your home can begin, [Builder] has several important tasks to accomplish that involve outside people and entities. For example:

- ❏ Structural changes you decided to make to the plans may necessitate revision of engineering for the home. This must be completed prior to applying for a building permit and can take from several days to several weeks.
- ❏ Residential construction requires that we obtain a building permit. The process varies and can take a few days to many weeks depending on the volume of applications being processed by the building department. This volume varies from month to month.
- ❏ The time of year may affect the start date because of the weather conditions.

Preconstruction Agenda

Purchasers _____ Date_____

Address _____

At the office:

- ☐ 1. Site plan
- ☐ 2. Soil report
- ☐ 3. Drainage plan
- ☐ 4. Status of permit
- ☐ 5. Utilities status
- ☐ 6. Homeowner association issues
- ☐ 7. Landscape plans
- ☐ 8. House plans
- ☐ 9. Specifications
- ☐ 10. Selections and options
- ☐ 11. Change orders
- ☐ 12. Change order cutoff schedule
- ☐ 13. Target start date
- ☐ 14. Construction sequence/schedule
- ☐ 15. Events that extend schedule
- ☐ 16. "Nothing's happening"
- ☐ 17. Quality, builder's inspection of work
- ☐ 18. Site visit guidelines
- ☐ 19. How to handle questions
- ☐ 20. Pre-drywall tour
- ☐ 21. Target deliver
- ☐ 22. Read maintenance and warranty

- ☐ 23. Other [Builder] topics:

- ☐ 24. Other client topics:

At the site:

- ☐ 25. Lot boundaries
- ☐ 26. Easements
- ☐ 27. Orientation of home
- ☐ 28. Trees and other natural features
- ☐ 29. Drainage
- ☐ 30. Mailbox location

Notes:

Purchaser _____ Date_____
Purchaser _____ Date_____
Builder _____ Date_____

[Builder] Homeowner Manual

Safety

We understand that you will want to visit your new home between these construction reviews. A new home construction site is exciting and can also be dangerous. Your safety is of prime importance to us. Therefore, we must require that you contact [Builder] before visiting your site. We reserve the right to require that you wear a hard hat and that a member of our staff accompany your during your visit. Please observe commonsense safety procedures at all times when visiting:

- Keep older children within view and younger children within reach, or make arrangements to leave them elsewhere when visiting the site.
- Do not walk backward, even one step. Look in the direction you are moving at all times.
- Watch for boards, cords, tools, nails, or construction materials that might cause tripping, puncture wounds, or other injury.
- Do not enter any level of a home that is not equipped with stairs and rails.
- Stay a minimum of six feet from all excavations.
- Give large, noisy grading equipment or delivery vehicles plenty of room. Assume that the driver can neither see nor hear you.

In addition to safety considerations, be aware of the possibility that mud, paint, drywall compound, and other construction materials are in use and can get onto your clothing.

Frame Tour

Many buyers appreciate the opportunity to tour their home just after the rough mechanical stage, before insulation. The rooms have begun to take shape but the inner workings are still visible. This is an opportunity for you to see the quality that goes inside the walls of your home.

Although this is not an opportunity to request changes, the meeting does give all of us an opportunity to confirm that we are correctly installing the options you ordered or approved changes you requested. We will also update you on the target delivery date during the frame tour.

As with the preconstruction conference, your frame tour is scheduled by your sales counselor. You will meet your builder at your new home. Frame tours usually take 20 to 30 minutes. Please remember to bring this homeowner manual, selection sheets, and any approved change orders.

Please understand that if for any reason you are unavailable to attend this meeting, we must continue with construction.

Locks and Keys

Once exterior doors and locks are installed, we will access your home with a construction master key. Company policy prohibits staff members from loaning these keys to customers. When you take possession, using your permanent key in the locks for the first time will reposition the lock tumblers and the construction master keys will no longer open your home.

Plans and Specifications

The building department of the city or county where your home is to be located must review and approve the plans and specifications for your home. We construct each home to comply with the plans and specifications approved by the applicable building department. Your specifications become part of our agreements with trade contractors and suppliers. Only written instructions from [Builder] can change these contracts. Many factors can cause variations between the model home you viewed and the home we deliver to you.

Regulatory Changes

From time to time, city or county agencies adopt new codes or regulations that can affect your home. Such changes are usually adopted in the interest of safety and are legal requirements with which [Builder] must comply. Therefore, builders may construct the same floor plan slightly differently in two different jurisdictions or at two different times within the same jurisdiction.

Individual Foundation Designs

Another area where variations among homes can appear is in the foundation system. The foundation design is specific to each lot. Based on the results of a soil test, an engineer determines which foundation system to use. Because of variations in soil conditions among lots, your foundation may differ from your neighbors' foundation or that of the same home in another neighborhood.

Topography and Homesite Conditions

Because each homesite is shaped differently, the position of your home on the site may vary from others in the community. You will receive a copy of a plot plan, a drawing that shows you the home's position on your homesite, at your preconstruction conference.

In addition, the exterior elevations of each home are affected by the topography, or surface contours, of your homesite. For instance, slope on the site may affect the number and configuration of the driveway, walks, steps, and rails. Exterior finish varies in accordance with the slope on the site and retaining walls are sometimes needed for extreme conditions. [Builder] identifies existing trees on your homesite that must be removed to create room for your home, drive, and so on. Our construction practices include steps intended to preserve other trees in a healthy condition. However, because the reaction of trees to construction activities and your new home are outside our control, we cannot guarantee the health or survival of any existing trees.

Utilities and Mailboxes

The location of meters, phone and electrical junction boxes, and mailboxes are examples of items outside the control of [Builder]. The authority of the utility companies and the U.S. Postal Service to designate the placement of these items is well established.

Changes in Materials, Products, and Methods

The new-home industry, building trades, and product manufacturers are continually working to improve methods and products. In addition, manufacturers sometimes make model changes that can impact the final product. For instance, appliance manufacturers generally make design changes every year. The model homes will show the appliances that were current when the models were built although your home may have a more recent version.

In all instances, as required by your purchase agreement, any substitution of method or product that we make will have equal or better quality than that shown in our models. Since such substitutions or changes may become necessary because of matters outside our control, we reserve the right to make them without notification.

Models

Model homes are equipped with larger capacity air conditioners to accommodate high traffic; models also display many decorator items, window coverings, and furnishings. Mature landscaping, extra walks, fences, lighting, fountains, signs and flags are other examples of items which are not part of the home we will be building for you. Please review your home's specifications as well as information [Builder] provides about optional items displayed in the models carefully to avoid misunderstandings. Contact your sales counselor with any questions.

Because finish sizes can vary somewhat, you should measure for window coverings in your home rather than in any model.

Televison and the Internet

You may be aware of various home construction methods and materials from watching television programs or exploring the Internet. [Builder] routinely reviews new approaches with a focus on building homes with materials and methods that perform predictably and to our standards. While we will be happy to discuss alternative methods and materials you may be interested in, we take a conservative approach to utilizing new approaches until they have been proven over time. In addition, what is appropriate for a home in one area may not be appropriate for your home because of soil, climate, and other conditions.

[Builder] Homeowner Manual

Natural Variations

Dozens of trade contractors have assemble your home. The same individuals rarely work on every home and, even if they did, each one would still be unique. The exact placement of switches, outlets, registers, and so on will vary slightly from the model and other homes of the same floor plan.

Quality

Our company will build your new home to the quality standards described in our documents and demonstrated in our model homes. Each new home is a handcrafted product—combining art, science, and raw labor. The efforts of many people with varying degrees of knowledge, experience, and skill come together.

Errors and Omissions

From time to time during a process that takes several months and involves dozens of people, an error or omission may occur. We have systems and procedures for inspecting our homes to ensure that the level of quality meets our requirements. We inspect every step of construction and are responsible for quality control. In addition, the county, city, or an engineer conducts a number of inspections at different stages of construction. Your home must pass each inspection before construction continues.

Your Questions

We also respect your interest and appreciate your attachment to the new home. Therefore, your input into our system is welcome. However, to avoid duplication of efforts, confusion, misunderstandings, or compounding errors, we ask that you first check your purchase documents to review what you ordered and the specifications for construction of your home. If you still believe we are in error, do one of two things:

1. Bring your concern up at the frame stage tour.
2. Contact your sales counselor, in writing, with your question. You are welcome to use one of the *Our Customer Wants to Know* forms included at the end of this section. We will note the date and time it was received and will respond within two business days. Also keep the following points in mind once you have notified the builder of a concern:
   - Your concern may involve a detail [Builder] has already noticed or appreciates your pointing out. Still, correction may not occur immediately. Trades and suppliers often impose trip charges for extra visits to the homesite so to be efficient, we may schedule the correction for the next routine visit. Also, a particular trade may be unavailable on short notice.
   - Work may simply be incomplete; an early stage can look wrong to you but be exactly right when finished.

[Builder] Homeowner Manual

- Methods and materials vary from region to region and change over time. When you are familiar with one method, you naturally question a different one. That does not make the new method wrong. Ask questions until you are comfortable.

Ugly Duckling Stages

During the construction process, every home being built experiences some days when it is not at its best. Homes under construction endure wind, rain, snow, foot traffic, and activities that generate noise, dust, and trash. Material scraps are a byproduct of the process. Although your new home is cleaned by each trade upon completion of their portion of the work, during your visits you will encounter some messy moments. Keep in mind that the completed homes you toured also once endured these "ugly duckling" stages.

Private Home Inspectors

If you wish to retain the services of a private home inspector to review your home during or at the end of construction, please be aware of [Builder] policies regarding private home inspectors. Your inspector:

- Must provide us with evidence of current worker's compensation and liability insurance.
- Should be a member of a professional association such as the American Society of Home Inspectors.
- Should be familiar with the codes applicable in your jurisdiction.
- Should be experienced with new home construction.
- Is responsible for staying informed as to the stage of construction the home has reached.
- Should avoid making any markings on the home itself.
- Should provide you and [Builder] with a written report of any concerns.

[Builder] will address concerns involving building code or contract issues only. Your sales counselor can provide you with a list of private home inspectors who have provided us with evidence of the required insurances as well as information about typical fees and services they offer.

Single Source

[Builder] is a single source company. That means that we select all personnel and companies who will contribute to your home. We order all materials and products from suppliers with whom we have established relationships. Although sweat equity arrangements are unavailable as a part of our purchase agreement, you are welcome to add your personal touches to the home after you close and take possession of it.

Trade Contractors

Your home is built through the combined efforts of specialists in many trades—from excavation and foundation, through framing, mechanicals, and insulation, to drywall, trim, and finish work. In order to ensure you the [Builder]'s standard of construction, only authorized suppliers, trade contractors, and [Builder] employees are permitted to perform work in your home.

Each trade contractor works on a limited portion of the home; they may not be aware of all the details that affect the home and are not in a position to offer judgments. All questions or requests for changes should go through [Builder], and we will obtain input from trades when that is appropriate.

Suppliers and trade contractors have no authority to enter into agreements for [Builder]. For your protection and theirs, the terms of our trade contractor agreements prohibit alterations without written authorization from [Builder]. Their failure to comply with this procedure can result in termination of their contract. Discuss changes you are considering with your sales counselor.

Schedules

The delivery date for your new home begins as an estimate. Until the roof is on and the structure is enclosed, weather can dramatically affect the delivery date. Even after the home itself is past the potential for weather-related delays, weather can severely impact installation of utility services, final grading, and concrete flatwork, to mention a few examples. Extended periods of wet weather or freezing temperatures bring work to a stop in the entire region. When favorable conditions return, the tradespeople go back to work, picking up where they left off. Please understand that they are as eager as you are to get caught up and to see progress on your home.

Delivery Date Updates

[Builder] recognizes that timing is critical to planning your move. Although a guaranteed date is unrealistic in the early stages of construction, the builder can provide regular updates. As the home nears completion, the builder can provide a firm delivery date (usually 45 days before the closing). Meanwhile, be flexible and avoid making arrangements that might cause you worry if the move-in date changes.

We will update you on the estimated delivery date at each of our construction meetings. You are also welcome to check with us for the most current target date. As completion nears, more factors come under our control and we can be more precise about that date. Expect a firm closing date no later than 30 days before delivery.

We suggest that, until you receive this commitment, you avoid finalizing arrangements for your move. Until then, flexibility is the key to comfort, sanity, and convenience. We want you to enjoy this process and avoid unnecessary stress caused by uncertainty that cannot be avoided. Review the Loan Lock heading in Section 3, Applying for Your Loan, for additional suggestions on this topic.

[Builder] Homeowner Manual

Please keep in mind that your belongings may be brought into the home only after the closing because of insurance issues and the regulations of the applicable building department.

"Nothing's Happening"

Expect several days during construction of your home when it appears that nothing is happening. This can occur for a number of reasons. Each trade is scheduled days or weeks in advance of the actual work. This period is referred to as "lead time." Time is allotted for completing each trade's work on your home. Sometimes, one trade completes its work a bit ahead of schedule. The next trade already has an assigned time slot, which usually cannot be changed on short notice.

Progress pauses while the home awaits building department inspections. This is also part of the normal sequence of the construction schedule and occurs at several points in every home. Also, throughout construction of a home, work progresses rapidly at some stages as highly visible stages are completed (such as installing large expanses of walls) and more slowly at others (such as detail work in framing in soffitts and closets). If you have questions about the pace of work, please contact our office for an update.

Construction Sequence

Although the specific sequence of construction steps varies and overlaps, generally we build your home in the following order:

Foundation
    Excavation
    Footer or caisson installation
    Form and pour walls
    Perimeter drain, if applicable
    Waterproof
    Insulation, if applicable
    Inspection
Framing
    First floor
    Second floor
    Roof trusses
    Roof sheathing
Roofing
    Felt or paper
    Valley flashing
    Shingles

Exterior

Exterior trim
    Fascia (boards at ends of rafters)
    Windows and doors
    Sheathing
    Finish materials
    Trim
    Deck, if applicable
    Gutters, if applicable
Exterior painting or staining
Concrete or asphalt
Fine grading
Landscaping, if applicable

Interior

Rough-in of mechanical systems
    HVAC (heating, ventilating, and air conditioning)
    Plumbing
    Electrical (extra outlets need to be installed at this point)
    Rough inspections
Insulation
Drywall
    Hang
    Inspection
    Tape and texture
Interior trim
    Doors
    Baseboards, casings, other details
Paint and stain
Finish work
    Cabinets
    Countertops
    Tile
    Floor coverings
    Appliances
    Hardware
    Screens
    Light fixtures
    Plumbing fixtures

[Builder] Homeowner Manual

Construction cleaning
Builder's punchlist
Improvement survey
Certificate of occupancy
Homeowner orientation
Closing
Home maintenance

Our Customer Wants to Know ...

Date _____ Lot # _____
Purchaser _____ Phone _____
Fax _____ E-mail _____

Question

_____
_____
_____
_____
_____
_____
_____

My preference is to receive a response by ___ phone ___ fax ___ e-mail ___ letter

Response

_____
_____
_____
_____
_____
_____
_____

☐ See attached

By _____

Date _____

Section 6: Homeowner Orientation

✔ Schedule–[Builder] sets orientation appointment Monday through Friday, between 8:00 A.M. and 3:00 P.M.; the meeting takes approximately 2 hours

✔ Last-Minute Activity–many items are fine-tuned in the last few days before delivery

✔ Preparation–hints on how to get the most from your orientation

✔ Completion of Items–many items will be completed prior to your move-in, and any remaining work will be performed by appointment

✔ *Feedback on Orientation Items*–an extra check and balance to ensure we have completed the work we committed to performing during your orientation

✔ *Orientation Forms*–copies of the orientation forms for your review; in particular, note the information regarding cosmetic surfaces on the first page of this set of forms

Homeowner Orientation

Your homeowner orientation is an introduction to your new home and its many features. We follow a preplanned agenda and a set route through the home to assure that we cover everything. Our homeowner orientation provides you with a

- Demonstration of your new home.
- Review of key points about maintenance and limited warranty coverage.
- Confirmation that [Builder] installed selections and options as you ordered them.

Scheduling

We schedule the orientation with you as your home nears completion, typically several days before your closing. Appointments are available Monday through Friday, 8:00 a.m. to 3:00 p.m. Especially in winter months, beginning by 3:00 P.M. assures sufficient day light to view all surfaces adequately. We meet at your new home. Expect your orientation to take approximately 2 hours.

Last-Minute Activity

If you visit your home a day or two prior to orientation, you may notice dozens of details that need attention. During the last few days just prior to your orientation appointment, many tradespeople and [Builder] employees will be working in your home. They are completing last-minute adjustments and fine-tuning your home. These finishing touches cannot be performed until all of the parts have been installed. What seems like a rush of activity is a normal part of the construction process.

Preparation

Following these hints will assure that you get the maximum benefit from your orientation.

Allow Enough Time

Arrange your schedule so you can use the full amount of time allotted.

Bring This Manual

By having this manual with your selection sheets and any approved change orders with you, any questions about the items installed in your home can usually be answered conveniently and immediately.

Attend Alone

Our experience shows that the orientation is most beneficial when buyers focus all their attention on their new home and the information we present. Although we appreciate that friends and relatives are eager to see your new home, it would be best if they visit after your orientation. Similarly, we suggest that, if possible, children and pets not accompany you at this time. If a real estate agent has helped you with your purchase, he or she is not required to attend.

Review Orientation Forms

We have included copies of our orientation forms at the end of this section. We note details that need attention on the orientation forms.

Cosmetic Surfaces

Cosmetic surface damage caused during construction is readily noticeable during the orientation. Such damage can also occur during the move-in process or through daily activities. Therefore, during your orientation, we will confirm that all surfaces are in good and acceptable condition. Any details that need attention will be listed on your orientation forms. After we correct any items noted during the orientation, repair of cosmetic surface damage is your responsibility. Additional details appear on the orientation forms.

Our limited warranty specifically excludes repairs for damage caused by moving in or living in the home. If your movers scratch the marble entry floor bringing the piano in, notify the moving company. If you splinter some wood trim and break a taillight backing out of your new garage, repairs to the garage and the car are your responsibility. [Builder] is always available to assist you with information about cosmetic repairs you may need to make.

Bring Questions

If you have not already done so, please read the maintenance information, limited warranty, and warranty guidelines in Section 8 of this manual. If you have questions, make note of them to bring up at the orientation.

Attire

Wear shoes that are convenient to get off and on. We will tour both the exterior and interior of your home. Anticipate that some dust, bending, kneeling, and reaching may be encountered.

Get Involved

Plan to listen carefully and take a hands-on approach. Push buttons, lock locks, and flip breakers. This helps you remember the dozens of details we cover.

Quality

The overall quality of your home should equal that shown in our models and described in your purchase documents. We list items we agree need further attention and arrange appropriate work. Orientation items fall into several categories:

- Incomplete or missing (Cabinet knob not installed.)
- Incorrect (Porch light should be polished brass, not antique.)
- Dysfunctional (Bath fan does not come on.)
- Below company standard (Mitered corner rough, top right of den door, hallway side.)
- Damaged (Scrape on wall from carpet installation.)
- Uncleaned (Mud on the garage floor.)

At some point, quality ceases to be scientific and becomes a matter of personal taste. In a few areas, your personal standards may be even higher than ours. Our commitment to you is that we will deliver what we promised. If you wish to make it even better after moving in, we will be happy to assist you with information.

Completion of Items

[Builder] takes responsibility for resolving any items noted. We will complete most items before your move-in. If work needs to be performed in your new home after your move-in, construction personnel are available for appointments Monday through Friday, 8:00 a.m. to 4:00 p.m.

Under normal circumstances, you can expect us to resolve all items within 15 working days. We will inform you of any delays caused by back-ordered materials. Please note that we will correct only those items listed. No verbal commitments of any kind will be honored by [Builder].

Gaining access to occupied homes to complete orientation items is a concern to homeowners and builders alike. [Builder] asks that you make appointments so that someone over 18 is present for repairs. Working around your busy schedule may result in service taking longer than anyone wants. Your cooperation is essential. Service hours are 8:00 a.m. to 4:00 p.m., Monday through Friday.

We will confirm that any items listed during your orientation have been resolved to meet our standards and policies. To be doubly certain of this, we mail you a written form for feedback. You will find a copy of this form on the next page.

Feedback on Orientation Items

Your satisfaction with your new home is important to us. Our records indicate that your Homeowner Orientation list has been completed. We would like your confirmation of that. A copy of that list is attached. Please review it and confirm that all items listed on it have been resolved.

If we have overlooked any detail from the original list, please note the number of the item in the space below. If all items have been satisfactorily resolved, simply sign the acknowledgment below. Either way, please return this form in the enclosed envelope by _____.

As always, your comments about our service or your new home are most welcome.

Please let us hear from you!

Sincerely,

Orientation Rep

....

___ All homeowner orientation items have been resolved.

___ The following homeowner orientation items still need attention (simply list the item numbers): _____

Comments: _____
_____
_____
_____

_____   _____
Homeowner   Phone

_____
Date

Orientation

Date_____ Lot #_____
Purchasers _____
Address _____
New Phone _____

We believe that your home is complete, in satisfactory condition, and meets the quality standards described in your contract documents. We invite your confirmation of this fact by offering you an opportunity to review your home at this time. Your signature indicates that with the exception of items noted on page 2, the components listed below are in good and acceptable condition, including, where applicable, the cosmetic surfaces of these items. Cosmetic damages noted subsequent to those identified today and listed on page 2 are excluded from warranty coverage except as specifically described in your homeowner manual.

Cosmetic surfaces are in acceptable condition

- ___ Appliances
- ___ Brass fixtures
- ___ Cabinets
- ___ Carpet
- ___ Caulking
- ___ Ceramic tile/grout (walls, counters, floors)
- ___ Countertops
- ___ Decks and exterior rails
- ___ Doors
- ___ Drywall
- ___ Fireplace doors
- ___ Garage overhead doors
- ___ Hardware (knobs, towel bars)
- ___ Hardwood floors
- ___ Landscaping (sod, shrubs, trees)
- ___ Light fixtures
- ___ Marble or manufactured marble
- ___ Masonry
- ___ Mirrors and medicine cabinets
- ___ Paint
- ___ Plumbing fixtures (sinks, tubs, faucets)
- ___ Resilient floor coverings
- ___ Shower or tub enclosure
- ___ Siding
- ___ Stair rail
- ___ Stucco
- ___ Windows, screens, patio doors
- ___ Wood trim

Selections/change orders

- ___ All selections and change order items are installed

Status summary

- ___ Grade: Complete Pending
- ___ A/C: Charged Pending N/A
- ___ Crawl: Dry Damp N/A
- ___ Smoke detectors respond to test buttons
- ___ GFCIs respond to test/reset buttons
- ___ Outside faucets function without leaks

Manufacturer literature/parts delivered

- ___ Heat system
- ___ Air conditioning
- ___ Humidifier
- ___ Water heater
- ___ Range
- ___ Cooktop
- ___ Range hood
- ___ Microwave
- ___ Dishwasher
- ___ Disposal
- ___ Fireplace
- ___ Broiler pan
- ___ Disposal wrench
- ___ Sink strainer and drain cover
- ___ Garage door openers or keys
- ___ Paint and stain samples

Warranty service

For your protection and to allow efficient operation of our services, our warranty system is based on your written list of items. Please refer to Section 8 of your homeowner manual for complete details.

Continued-

Date _____   Lot # _____
Purchasers _____
Address _____

Inspection Items	**Company Use**

Homeowner _____   Builder _____

....

All items listed above have been resolved.

Homeowner _____   Builder _____

Date _____   Date _____

Note to Home Buyer:

At the end of your homeowner orientation, you will receive:

- A list of emergency phone numbers for critical trade contractors, such as heating and plumbing, who might be needed after hours or on weekends. We suggest you insert these phone numbers at the front of Section 8, Caring for Your Home, so that you can find them quickly in an emergency.

- The manufacturer's literature for the furnace, water heater, and other consumer products. Copies of this material for standard items are available for your review in our sales office.

- Copies of completed orientation forms. We suggest you insert those forms here.

Section 7: Closing on Your Home

- ✔ Date–[Builder] provides a minimum of 30 days' notice for the closing

- ✔ Location–we confirm the location of your closing appointment when we set the appointment

- ✔ Documents–an overview of the materials that you will sign at closing

- ✔ "The Final Number"–due to prorations that are based on your closing date, the final amount you will need to bring can be determined only after your closing date is set

- ✔ Preparation–reminders to assure you have addressed all necessary tasks prior to closing

- ✔ House Keys–delivered to you at the closing table

- ✔ Garage Door Opener Operators–left in a kitchen drawer in your new home

- ✔ Mailbox Keys–available from your post office upon presentation of your personal identification

- ✔ First Mortgage Payment–your lender will inform you where to send your house payments and when the first payment will be due

- ✔ Storing Documents–your closing documents are valuable papers; store them safely

- ✔ Utility and Community Services–names and numbers for your convenience

- ✔ Moving Hints–some reminders and checklists to make moving easier

Closing on Your Home

At closing the ultimate purpose of your purchase agreement is achieved: Ownership of your new home is transferred from [Builder] to you. The steps include finalizing your loan (one set of papers and checks) and [Builder] selling you the home (another set of papers and checks). The funds are disbursed to the appropriate people and companies, title is transferred to your name(s), and the loan is recorded against your new property. This process involves about 75 documents–some of which are duplicates. Although these documents are not negotiable and thousands of home buyers have signed them, you should read them.

Date

[Builder] recognizes that timing is vitally important in planning your move and locking in your loan. We can specify an exact delivery date when construction reaches a point at which weather, material and labor shortages, lender issues, or change orders are unlikely to affect completion of your home. The closing takes place shortly after your orientation. [Builder] will notify you of the closing date 30 days or more before the date. We set the specific appointment time with at least 3 day's notice. Typically, the closing process takes about an hour.

Location

The closing on your new home typically takes place at the title company, although it occasionally occurs at the lender's office. We confirm the location with you when we set the appointment.

Documents

At closing, the documents necessary to convey your new home to you and to close the loan from the mortgage company will be signed and delivered. In addition to these standard items, the lender, the title company, and [Builder] may require other documents to be signed. The principal documents typically include the following:

General Warranty Deed

The general warranty deed conveys the home and lot to you, subject only to permitted exceptions. This does not apply if you already own the lot.

Title Commitment

At or before closing, we will deliver to you a standard form for an American Land Title Association (ALTA) owner's title insurance commitment to insure salable title of your home to you in the amount of the purchase price, subject to the permitted title exceptions that may be described in the purchase agreement.

The title insurance company will mail the actual policy in the weeks following the closing. When you receive this, keep it in a safe place with your other important papers. What you will see on the day of closing is a document that promises to issue the policy. Lenders require title insurance in the amount of the mortgage. This insurance protects the lender in the event the title search missed anything. You are wise to request an owner's policy to protect your interest in the property. By ordering the owner's policy from the same company that issues the lender's policy, you can save a bit; the title insurance company will usually issue a second policy at a discount.

Review the title commitment carefully. Discuss any questions with your title company. Within 60 days after the closing, the title company mails a standard ALTA owner's title insurance policy, insuring you the title to your home in accordance with the commitment you received at closing. Keep the title insurance policy with your other valuable papers.

[Builder] Limited Warranty

We provide a copy of the limited warranty in this manual for your review. Please read it thoroughly.

Promissory Note

The promissory note is from you, payable to the lender in the principal amount of the loan, plus interest. One-twelfth of your annual taxes and homeowner's insurance will be added to the principal and interest payment to determine your total monthly payment.

Deed of Trust

This encumbers your home as security for repayment of the promissory note.

Homeowner Association Documents

You will receive and sign for another copy of your homeowner association covenants, conditions, and restrictions; the association bylaws; and articles of incorporation at closing. [Builder] recommends that you read these carefully. The provisions they contain will be enforced.

"The Final Number"

Certain customary items in connection with the property will be prorated to the date of closing such as prepaid expenses, or reserves required by your lender and homeowners association, if applicable. Prorations of general real property taxes and assessments will be based on the current year's taxes and assessments or, if they are unavailable, on the taxes and assessments for the prior year.

The final cost figure is available near to the actual closing. Although a reasonably close estimate may be determined before the date of closing, the proration of several items included is affected by the closing date and cannot be calculated until that date is known. The Real Estate Settlement Procedures Act (RESPA) provides you with many protections. Under this law, you can review the settlement page that lists costs you are paying at closing one day before the closing appointment.

Preparation

The key to a smooth closing is preparation. Several details require your attention. You can handle most of these by phone. Address these details during the weeks before closing to prevent last-minute delays.

Form of Payment

Plan to bring cash, certified funds, or a bank check (made out to yourself, which you will endorse at the closing) to the closing table. In your planning, be sure to allow time to arrange for and obtain these funds. Keep in mind that some banks place a hold on monies moved from another account.

Insurance

You need to provide proof of a homeowner's policy from your insurance company. Your insurance agent should know exactly what is needed. We suggest you arrange for this at least 3 weeks before the expected closing date.

[Builder] or Lender Issues

The title company is not authorized to negotiate or make representations on behalf of any of the parties involved in the closing. Therefore, please discuss any questions, agreements, or other details directly with us or your lender in advance of the closing.

[Builder] Homeowner Manual

Utilities

[Builder] will have utility service removed from its name 3 days after closing. You will need to notify all applicable utility companies of your move so that service is provided in your name. We suggest that you contact these companies well ahead of time to avoid any interruption in service. If you ordered a security system for your home, you can arrange to activate that system by contacting the monitoring service for a connection appointment. For your convenience, we have included a list of your utility companies and contact information at the end of this section.

House Keys

When the closing process is complete, you will get the keys to your new home. You will receive two keys for each lock on your home. The same key will operate both the knob and the deadbolt locks. When you insert your new key for the first time in each lock, the tumblers are altered and our master key will no longer unlock your door.

We recommend that you try all of the keys in all of your locks to confirm smooth operations. Depending on the number of family members living in the home, you may want to get extra copies of your house keys made.

Garage Door Opener Operators

Garage door opener operators, if applicable, will be left in a drawer in your kitchen. If you wish to change the code, review the manufacturer instructions. Batteries typically need to be replaced about once a year. You will receive two operators for each garage door opener installed in your home. If you need additional operators, contact the garage door opener company using the customer service number shown in the manufacturer's literature that came with the openers.

Mailbox Keys

U.S. Postal Service regulations state that, [Builder] is not permitted to deliver mailbox keys to you. Mailbox keys, where applicable, are available from your post office. You will need proof of identity, and you will be asked to sign for your keys.

First Mortgage Payment

Your lender will provide you with information on where to send your mortgage payments and when the first payment will be due. Many lenders supply payment coupons for you to send in with your payments.

Storing Documents

We suggest that you store the legal documents from your closing with other valuable papers, in a safe place. You will need them for tax purposes and when you refinance or sell your home.

Utility and Community Services

Gas
<Name>
<Address>
<Phone>
<Fax>
<Hours>

Notes

Electric
<Name>
<Address>
<Phone>
<Fax>
<Hours>

Notes

Telephone
<Name>
<Address>
<Phone>
<Fax>
<Hours>

Notes

Recycling
<Name>
<Address>
<Phone>
<Fax>
<Hours>

Notes

Water
<Name>
<Address>
<Phone>
<Fax>
<Hours>

Notes

Sewer
<Name>
<Address>
<Phone>
<Fax>
<Hours>

Notes

Trash Collection
<Name>
<Address>
<Phone>
<Fax>
<Hours>

Notes

Post Office
<Name>
<Address>
<Phone>
<Fax>
<Hours>

Notes

Cable TV
<Name>
<Address>
<Phone>
<Fax>
<Hours>

Notes

Newspaper
<Name>
<Address>
<Phone>
<Fax>
<Hours>

Notes

Moving Hints

Take precautions to protect vulnerable surfaces such as hardwood or resilient floors. Cover rails with moving pads or blankets. Remove doors where furniture might be a tight fit. You can protect carpet with ribbed, plastic runners.

Professional movers should have insurance for any damage they might accidentally cause. Friends and relatives will not. They are also unlikely to have the training and practiced skills of professional movers. If you are moving yourself, organize the schedule to avoid rushing and include rest breaks. People who are tired or in a hurry are more likely to hurt themselves or your belongings.

Whatever else is going on, at dinner time assemble the family for your first meal together in the new home. Sit across the card table from each other, smile, and say "We made it."

Moving Preparation Checklist

- Compare proposals of professional movers:

    –Costs for services such as packing and unpacking
    –Costs of packing materials and boxes
    –Distance and weight charges
    –Insurance
    –Availability and notice needed

- Plan a self-move well in advance:

    –Make truck reservation early (6–8 weeks, or more)
    –Include a reservation for a dolly and moving pads
    –Reconfirm one week prior

- If you have children, involve them in planning and preparing for the move
- Create a file for storing documents about your home and manufacturer literature
- Retain receipts for tax purposes. Moving costs may be deductible
- Send change-of-address cards to magazines and book clubs six weeks prior to your move
- Give the forwarding order to your old post office one month prior to assure uninterrupted service
- Register children in their new schools
- Transfer medical and dental records, if necessary
- Arrange for homeowner insurance and obtain the certificate you need for closing
- Order checks with new your address; update financial records
- Update your driver's license, car and voter registration
- Properly dispose of flammable or hazardous materials that should not be moved

[Builder] Homeowner Manual

- **Packing Materials**

  - Boxes of various sizes; cartons for mattresses
  - Packing tape and heavy string
  - Packing paper, newspaper, bubble wrap
  - Labels to identify boxes (include a number, room/name); "Fragile" labels for special items
  - Markers
  - Master packing list (list each box by number with name/room and brief description of contents)
  - Scissors
  - Furniture pads, blankets, rugs

- **Moving Day Necessities**

  - Children's toys and games
  - Toilet paper
  - Beverages and snacks
  - Paper towels
  - Soap and hand towels
  - Trash bags
  - First aid kit
  - Prescription medication
  - Medical supplies for special needs
  - Pad and pen
  - Shelf liners
  - Small tools: Tape measure, scissors, screwdrivers, hammer
  - Ice maker hook-up kit
  - Dryer vent flex hose
  - New hoses for washing machine
  - Picture hangers
  - Plant hooks
  - Scratch cover
  - Phone and phone book

[Builder] Homeowner Manual

Section 8: Caring for Your Home

✔ Homeowner Use and Maintenance Guidelines—introduction to the maintenance information in this manual

✔ [Builder] Limited Warranty Guidelines—introduction to the criteria [Builder] uses to screen warranty items

✔ Warranty Reporting Procedures—standard, emergency, miscellaneous, and appliance warranty procedures

✔ Warranty Item Processing Procedures—a simple description of a complex process

✔ Help Us to Serve You—things you need to know so we can provide effective warranty service

✔ Warranty Service Summary—a one-page guide to who to contact in various service situations

✔ Fire Prevention—reminders to prevent fire in your home

✔ Extended Absences—tips for preparing and reminders for the day you leave

✔ Energy and Water Conservation—suggestions consuming energy and water wisely

✔ Appliance Service—a worksheet where you can record serial and model numbers along with manufacturer service phone numbers

✔ Home Care Supplies—create a shopping list of and supplies you will need to care for your home

✔ Maintenance Schedule—a place to make notes about routine maintenance tasks and plan your schedule

✔ Air Conditioning through Wood Trim—an alphabetical list of the items in your home, including maintenance hints, warranty criteria, and troubleshooting tips

✔ Forms—for your convenience when reporting warranty items and giving us feedback about this manual

[Builder] Homeowner Manual

Caring for Your Home

[Builder] has constructed your home with carefully selected materials and the effort of experienced craftsmen and laborers under the supervision of our field personnel, with the administrative support of our office personnel. Although this group works from detailed plans and specifications, no two homes are exactly alike. Each one is unique; a home is one of the last hand-built products left in the world. Over time, each behaves differently.

Although quality materials and workmanship have been used in creating your home, similar to an automobile, it requires care from the first day. Regular homeowner maintenance is essential to providing a quality home for a lifetime. This section of our manual was assembled in to assist you in that effort.

Homeowner Use and Maintenance Guidelines

We are proud of the homes we build and the neighborhoods in which we build them. We strive to create lasting value. This is best achieved when you, as the homeowner, know and perform appropriate maintenance tasks. Periodic maintenance is necessary because of normal wear and tear, the inherent characteristics of the materials used in your home, and normal service required by the mechanical systems. Natural fluctuations in temperature and humidity also affect your home, resulting in maintenance items. The natural and manufactured materials, the components interact with each other and the environment.

We recognize that it is impossible to anticipate and describe every attention needed for good home care. We focused on items that homeowners commonly ask about. The subjects are listed in alphabetical order to make finding answers to your questions convenient. Because we offer home buyers a variety of floor plans and optional features, this manual may discuss components that are not present in your home.

Checklists

You will find several checklists included in this manual. These cover fire prevention reminders, energy and water conservation tips, suggestions for extended absences, appliance service information, home maintenance supplies list, and a maintenance schedule. Again we make no claim that we have included every detail. We do believe we have provided you with a good start, and we've allowed space for you to add your own notes to our checklists.

Prompt Attention

In addition to routine care, many times a minor maintenance attention provided immediately saves you a more serious, time-consuming, and sometimes costly repair later. Note also that neglecting routine maintenance can void applicable limited warranty coverage on all or part of your home.

[Builder] Homeowner Manual

By caring for your new home attentively, you ensure uninterrupted warranty coverage as well as your enjoyment of it for years. The attention provided by each homeowner contributes significantly to the overall desirability of the community.

Manufacturer Literature

Please take time to read the literature (warranties and use and care guides) provided by the manufacturers of consumer products and other items in your home. The information contained in that material is not repeated here. Although much of the information may be familiar to you, some points may differ significantly from homes you have had in the past.

We make every effort to keep the information in this manual current. However, if any detail in our discussion conflicts with the manufacturer's recommendations, you should follow the manufacturer's recommendations.

Activate specific manufacturer's warranties by completing and mailing any registration cards included with their materials. In some cases, manufacturer's warranties may extend beyond the first year and it is in your best interests to know about such coverages.

[Builder] Limited Warranty Guidelines

While we strive to build a defect-free home, we are realistic enough to know that, with repeated use, an item in the home may fail to perform as it should. When this occurs, we will make necessary corrections so the item meets our warranty guidelines. In support of this commitment, [Builder] provides you with a limited warranty.

Corrective Actions

In addition to the information contained in the limited warranty itself, this manual includes details about the criteria we will use to evaluate concerns you report. The purpose is to let you know what our warranty commitment is for the typical concerns that can come up in a new home. The manual describes the corrective action we will take in a many common situations.

We Sometimes Break Our Own Rules–In Your Favor

Our criteria for qualifying warranty repairs are based on typical industry practices in our region and meet or exceed those practices. Please note that we reserve the right, at our discretion, to exceed these guidelines if common sense or individual circumstances make that appropriate, without being obligated to exceed all guidelines to a similar degree or for other homeowners whose circumstances are different.

We Sometimes Say No

With a product as complex as a home, different viewpoints regarding which tasks are homeowner maintenance responsibilities and which are [Builder] warranty responsibilities are possible. If you request warranty service on a maintenance item, we will explain to you the steps you should take to care for the item. We are available to answer your home-care questions during and after your warranty period. Providing normal maintenance for your home is your job.

Warranty Specimen Provided for Your Review

You will receive the signed limited warranty document at your closing. We provide a specimen copy for your review at the time you sign your purchase agreement. Please read through this information, as well as the service procedures and guidelines discussed on the following pages. If you have any questions, please contact our warranty office.

Warranty Reporting Procedures

Providing warranty service for a new home is more complicated than for other products. When you purchased your home, your actually purchased hundreds of items and the work of 35 to 50 independent trade contractors. With so many details and people involved, a planned system is essential.

Our warranty service system is designed based on your written report of nonemergency items. This provides you with the maximum protection and allows us to operate efficiently, thereby providing faster service to all homeowners. Emergency reports are the only service requests we accept by phone. Please put all nonemergency service requests in writing.

You are welcome to mail, fax, e-mail, or drop off your list in person at our main office. Keep a copy for your records. This written system permits [Builder] personnel to focus their time producing results for you and following up. Experience has taught us that accuracy and efficiency suffer when we work outside this system and sacrifice careful documentation.

We plan two standard warranty contacts with you. The first is 60 days after your closing and the second is at 11 months after closing. We also have emergency response procedures and have provided for miscellaneous warranty requests between the standard 60-day and 11-month reports. Service for your appliances is handled differently and is described in detail on the next page.

60-Day Report

For your convenience and in order for our service program to operate at maximum efficiency, we suggest that you wait 60 days before submitting a warranty list. This allows you sufficient time to become settled in your new home and to use most components repeatedly. As you notice items, jot them down on a service request form (found at the end of this manual).

11-Month Report

Near the end of the eleventh month of your materials and workmanship warranty, you should submit a year-end report if you have any items to report. We will also be happy to discuss any maintenance questions you may have at that time. Again, keep notations of items on a service request form. This is also the best time for you to request the "one-time" repairs we offer on several components such as drywall.

Emergency Service

While emergency warranty situations are rare, when they occur, prompt response is essential. Begin by checking items you can check. Troubleshooting tips appear in this manual for several of your home's components:

- Air conditioning
- Electrical
- Heat system
- Plumbing
- Roof (leak)
- Water heater

Please refer to the individual categories to review these hints; you will find them at the end of the corresponding sections. Often the appropriate action by you can solve a problem immediately or mitigate the situation until a technician arrives.

If your review of the troubleshooting tips fails to solve the problem, during business hours, call [Builder]'s warranty office:

(555) 555-5555, ext. 555

After hours, or on weekends or holidays, call the necessary trade contractor or utility company directly. Their phone numbers are listed on the Emergency Phone Numbers sheet you receive at orientation. We suggest that you insert the Emergency Phone Numbers sheet in this section of your homeowner manual or secure it inside a kitchen cabinet, near your phone.

Our trade contractors or local utility companies provide emergency responses to the following conditions:

- Total loss of heat when the outside temperature is below 50 degrees F
- Total loss of electricity
- Total loss of water
- Plumbing leak that requires the entire water supply to be shut off
- Gas leak

Note that if a service (gas, electricity, water) is out in an entire area, attention from the local utility company is needed. Trade contractors are unable to help with such outages.

Air Conditioning. Understandably, if your air conditioner is not working, you want it fixed pronto. In a typical scenario, many other homeowners across our region will discover they too need service on their air conditioners on the same hot day that you do. The trades who address these needs generally respond to calls on a first come, first served basis. If your call for service comes during this time period, you may wait several days for a technician to arrive. For this reason, we recommend that you operate your air conditioner as soon as warm temperatures begin. In this way, if service is needed, you can avoid the rush and get a more satisfactory response.

Roof Leak. While we agree with homeowners that a roof leak is indeed an emergency, the reality is that repairs cannot safely or effectively be performed while the roof is wet. During business hours, contact our office with the information, take appropriate steps to mitigate damage, and we will follow up when conditions make repairs possible. (See *Roof* for more details.)

Other Emergencies. In addition to emergency situations covered by our limited warranty, be prepared for other kinds of emergencies. Post phone numbers for the fire department, police, paramedics, and poison control near phones in your home. Have companies in mind in the event you need a locksmith, water extraction, glass breakage repair, or sewer router service. If you are new to the area, neighbors may be able to recommend good service providers. Introduce your children to neighbors who might be available to help in an emergency if you are not home.

Other Warranty Service

If you wish to initiate nonemergency warranty service between the standard 60-day and 11-month report, you are welcome to do so by sending in a service request form (we've included 3 copies of this form at the back of this manual) or simply by writing a letter that includes your name, address, phone numbers, and a list of your concerns.

Homeowners who want to arrange 60-day or 11-month warranty visits receive priority scheduling. We schedule appointments for miscellaneous requests on a first come, first served basis between the standard appointments. As a result, service on miscellaneous requests may take a bit longer to address.

Kitchen Appliance Warranties

The manufacturers of kitchen appliances have asked to work directly with homeowners if any repairs are needed for their products. Customer service phone numbers are listed in the use and care materials for each appliance. Be prepared to provide the model and serial number of the item and the closing date on your home. For your convenience, we have included an Appliance Service information sheet among the other checklists in this manual.

Appliance warranties are generally for one year; refer to the literature provided by the manufacturer for complete information. Remember to mail in any registration cards you receive with manufacturer materials. Being in the manufacturer's system assures that in the event of a recall the company can contact you and arrange to provide the needed correction.

Warranty Item Processing Procedures

When we receive a warranty service request, we may contact you for an inspection appointment. Warranty inspection appointments are available Monday through Friday, 7:00 a.m. to 4:00 p.m. We inspect the items listed in your written request to confirm warranty coverage and determine appropriate action. Generally, reported items fall into one of three categories:

- Trade contractor item
- In-house item
- Home maintenance item

If a trade contractor or an in-house employee is required to perform repairs, we issue a warranty work order describing the situation to be addressed. If the item is home maintenance, we will review the maintenance steps with you and offer whatever informational assistance we can. Occasionally the inspection step is unnecessary. In that case, we issue the needed work orders and notify you that we have done so.

Help Us to Serve You

We can provide service faster and more accurately if we have all the necessary information. With your warranty request, please include:

- Your name, address, and the phone numbers where you can be reached during business hours.
- A complete description of the problem, for example, "guest bath—cold water line leaks under sink," rather than "plumbing problem."
- Information about your availability or the best days or times to reach you. For instance, if calling you at work is acceptable, let us know. Otherwise, we will use your home phone number. If you are usually home on Thursday, mention that.

Access to Your Home

[Builder] conducts inspections of interior warranty items only when an adult is available to accompany our representative and point out the items you have listed. Both our in-house service technicians and those of our trades contractors will likewise perform repairs only when an adult is available to admit them to your home. An adult is a person 18 or older who has your authorization to admit service personnel and sign completed work orders.

We do not accept keys, nor will we permit our trade contractors to accept your key and work in your home without an adult present. While we recognize that this means processing warranty service items may take longer, we believe your peace of mind and security should be our first concern.

Exterior Items

Exterior items can usually be inspected and repaired without an adult present, provided access is available (for instance, no locked gate). However, we will contact you the day prior to any visit and let you know we will have someone on your property. If you prefer to meet with us and discuss the item(s) in question, we are happy to arrange an appointment to do that.

Repair Appointments

Depending on the work needed, at the conclusion of the inspection appointment, the warranty manager will most likely ask you to designate a *work date*—a date a minimum of 10 days from the inspection date—for approved repairs to be made. This 10-day time frame allows us to notify appropriate trades people and arrange for most repairs to occur on the same day.

Although on occasion work must occur in sequence and more than one work date might be needed, this system works well in the majority of situations. Once work date appointments are set, we confirm them the day before and our warranty manager follows up to confirm repairs are completed.

Inspection and Work Hours

Many homeowners ask whether evening and weekend appointment times are available. [Builder] understands the desire for appointments outside normal business hours. We recognize the trend to services being available "24/7" in many businesses. However, in investigating how such appointments could be arranged, we discovered many factors that make extended service hours impractical.

- A significant portion of repairs require daylight for proper execution. This applies to drywall, paint, and exterior work of almost any type.
- We also found that most of the 35 to 50 independent trade contractors who helped us build your home—many of whom operate as small companies—were unable to work all week and also be available for extended hours. Therefore, the few repairs that could be performed in off-hours failed to eliminate the need for repair appointments during normal hours.
- Administrative staff and supervisors would need to be available to answer questions. Having some personnel work extended hours meant being short staffed during normal business hours.
- When we calculated the impact on wages and salaries for adding more personnel or compensating existing personnel for working non-traditional hours, we found that this affected overhead, and consequently the prices for our homes.

[Builder] Homeowner Manual

We are still looking for a workable long-term answer to this recognized dilemma. Meanwhile, our warranty hours will be as follows:

- Administrative staff: Monday through Friday, 8:00 a.m. until 5:00 p.m.
- Inspection appointments: Monday through Friday, 7:00 a.m. until 4:00 p.m.
- Work appointments: Monday through Friday, 7:00 a.m. until 4:00 p.m.

Evening and weekend appointments are reserved for emergency situations. We appreciate your understanding and cooperation with these policies.

Pets

[Builder] respects the pets that many homeowners count as members of their households. To prevent the possibility of an animal getting injured or lost, or giving in to its natural curiosity about tools and materials used for repairs, we ask that you restrict all animals to a comfortable location during any warranty visit, whether for inspection or warranty work. This policy is also for the protection of our employees and trades personnel. We have instructed [Builder] and trades personnel to reschedule the appointment if pets have access to the work area.

Your Belongings

In all work that we perform for our homeowners we are concerned that their personal belongings be protected. When warranty work is needed in your home, we ask that you remove vulnerable items or items that might make performing the repair difficult. [Builder] and trade personnel will reschedule the repair appointment rather than risk damaging your belongings.

Surfaces

We expect all personnel who work in your home to arrive with appropriate materials to cover the work area, protecting it from damage and catching the dust or scraps from the work being performed. Similarly, all personnel should clean up the work area, removing whatever excess materials they brought in.

Repair personnel will routinely check the work area for any existing damage to surfaces. They will document any scratches, chips, or other cosmetic damage prior to beginning repairs to avoid any later disagreement about how and when such damage occurred.

Signatures on Work Orders

Signing a work order acknowledges that a technician worked in your home on the date shown and with regard to the items listed. It does not negate any of your rights under the warranty nor does it release us from any confirmed warranty obligation. If you prefer not to sign the work order, the technician will note that, sign the work order and return it to us for our records. Our work order form includes a brief survey about the service provided. We appreciate your taking a moment to respond to the items listed and let us know your opinion. If you are dissatisfied with any service we provide, you can note that on the work order or call the warranty office with your feedback. We will review your concerns and determine whether our requirements have been met. While complaints of this type are infrequent, about 50 percent of the time we find the homeowner is correct and more attention is needed.

Completion Time

Regular review of outstanding work orders is part of our office routine. Checking with trades and homeowners alike, we strive to identify the cause for delays and get all warranty work completed within an appropriate and reasonable amount of time.

We intend to complete warranty work orders within 15 work days of the inspection unless you are unavailable for access. If a back-ordered part or similar circumstance causes a delay, we will let you know. Likewise, when weather conditions prevent the timely completion of exterior items, we track those items and follow up to ensure that they are addressed when conditions are right. This can mean a wait of several months.

Missed Appointments

Good communication is one key to successful completion of warranty items. We strive to keep homeowners informed and to protect them from inconvenience. One of our challenges in this regard is when unexpected events sometimes result in missed appointments.

If a [Builder] employee or a trade person will be late, he or she should contact you as soon as the delay is recognized, offering you a choice of a later time the same day or a completely different appointment. If you must miss an appointment, we appreciate being alerted as soon as you realize your schedule has changed. We can put work orders on "hold" for 10 to 30 days and re-activate them when your schedule offers a better opportunity to arrange access to the home.

[Builder] Homeowner Manual

Warranty Service Summary

The many details of warranty coverage can be confusing. We hope this summary of key points will help. If you do not know whom to contact, call our warranty office and we will guide you.

Warranty Hours

- Administrative staff: Monday through Friday, 8:00 a.m. until 5:00 p.m.
- Inspection appointments: Monday through Friday, 7:00 a.m. until 4:00 p.m.
- Work appointments: Monday through Friday, 7:00 a.m. until 4:00 p.m.

Appliances

Contact the manufacturer directly with model and serial number, closing date, and description of problem. Refer to your Appliance Service information sheet.

Emergency

First, check the troubleshooting tips under several individual headings in this manual. If those tips do not solve the problem, during our business hours (Monday through Friday, 8:00 a.m. until 5:00 p.m.), call our warranty office:

(555) 555-5555, ext. 555

After business hours or on weekends or holidays, contact the trade or appropriate utility company directly using the emergency numbers you receive at your orientation.

Nonemergency

Mail, fax, e-mail, or drop off your list of items at our warranty office. You will find warranty service request forms at the end of this manual or you can request copies by calling our office.

Phone (555) 555-5555, ext.555   [Builder]
Fax (555) 555-5555   1234 Some Street, Suite #
<e-mail>   City, State, Zip Code

Storm Damage or Other Natural Disaster

Contact your homeowner's insurance agent immediately. Contain damage as much as possible without endangering yourself. In extreme situations, photograph the damage.

Fire Prevention

Fire safety should be practiced by all family members. Awareness of potential dangers and preventive actions are preferable to even the fastest response. Keep these hints in mind and add your own reminders in the space provided on the next page.

Train Family Members

- Ensure that all family members know what escape routes exist in your home.
- Conduct a fire drill with family members.
- Test the smoke detectors to assure they function and so that everyone recognizes the sound. Follow the manufacturer's directions for cleaning and servicing all of your smoke detectors.
- As soon as possible, teach young children how and why to dial 911.
- Have a general use fire extinguisher and instruct all family members in its location and use.
- Teach children the safe use of appliances such as irons and toasters.

Practice Prevention

- Store matches away from children and heat sources.
- Avoid smoking in bed.
- Avoid leaving small children home alone, even for a short time.
- Maintain appliances in clean and safe working condition.
- Avoid overloading electrical outlets.
- Ensure that all electrical cords are in good repair.
- Use correctly sized fuses.
- Avoid having any flammable objects or materials near the stove.
- Keep the range hood filter clean to prevent a build up of grease.
- Allow space for cooling around electrical equipment.

[Builder] Homeowner Manual

❏ Unplug the iron when it is not in use. Do not leave an iron that is on unattended.

❏ Use electric blankets with care, following manufacturer directions.

❏ Store volatile materials (paint, gasoline for the lawn mower, and so on) in appropriate containers, away from flames (such as pilots lights) or heat sources. Many trash collection services offer a means for you to dispose of hazardous items. Check with your service provider for details.

❏ Keep the barbeque clear of flammable objects and materials.

❏ If your home includes a gas fireplace follow all directions and do not leave the fireplace unattended when it is on. If you have a wood burning fireplace:

- Arrange for professional cleaning of the chimney at appropriate intervals.
- Maintain the spark arrester on the chimney.
- Never use liquid fire starters (such as for a charcoal barbeque) in an indoor fireplace.
- Use a screen or glass doors when a fire is burning.
- Confirm the fire is out before closing the flue.
- Do not leave the fireplace unattended while a fire is burning.

❏ During holidays, ensure that all cords and connections are in good condition and of appropriate capacity for electrical decorations.

❏ If you decide to remodel, finish the basement, or add onto your home, obtain a building permit and work with trained professionals. Ensure that all building department inspections occur and that the work complies with all applicable codes. This also applies to installing a gas line for an outdoor barbeque, a gas fireplace, clothes dryer, and so on.

Your Additional Reminders and Notes:

_____
_____
_____
_____
_____
_____
_____
_____

Extended Absences

Whether for a vacation, business travel, or other reasons, nearly all of us occasionally leave our homes for days or weeks at a time. With some preparation, such absences can be managed uneventfully. Keep these guidelines in mind and add additional reminders that are appropriate to your situation.

Plan in Advance

- Ask a neighbor to keep an eye on the property. If possible, provide them with a way to reach you while your are away.

- If you will be gone an especially long time (over two weeks) consider arranging for a house sitter.

- Arrange for someone to mow the lawn or shovel snow.

- Notify local security personnel or police of the dates you will be away.

- Stop mail, newspapers, and other deliveries.

- Use lighting timers (available at hardware stores for $10 to $20).

- Confirm that all insurance policies that cover your property and belongings are current and provide sufficient coverage.

- Mark valuable items with identifying information. Consider whether you have irreplaceable items that should be stored in a bank vault or security box.

As You Leave

- Forward phone calls to a relative or close friend.

- Unplug computers and other electronic devices that might be harmed in an electric storm.

- Leave window coverings in their most typical positions.

- Confirm that all doors and windows are locked and the deadbolts are engaged.

- Shut off the main water supply. Set the thermostat on the water heater to "vacation" to save energy.

- Store items such as your lawn mower, bicycles, or ladders in the garage.

[Builder] Homeowner Manual

❏ Disengage the garage door opener (pull on the rope that hangs from the mechanism). Use the manufacturer's lock to bolt the overhead door. *Caution:* Attempting to operate the garage door opener when the manufacturer's lock is bolted will burn out the motor of your opener. Upon your return, unlock the garage door first, then re-engage the motor (simply push the button to operate the opener and it will reconnect) to restore normal operation.

❏ Leave a second car in the drive.

❏ Summer: Turn your air conditioner fan to on. Set the thermostat to 78.

❏ Winter: Set the thermostat to a minimum of 55. Leave doors on cabinets that contain plumbing lines open. Leave room doors open as well. This allows heat to circulate.

❏ Arm your security system, if applicable.

Your Additional Reminders and Notes:

[Builder] Homeowner Manual

Energy and Water Conservation

Good planning and thoughtful everyday habits can save significant amounts of energy and water. In the process of conserving, you also save money as an additional benefit. Keep these hints in mind as you select and use your home's features:

Heating and Cooling

❑ Maintain all your home's systems in clean and good working order to prevent inconvenience and maximize efficiency. Arrange for a professional to service heat and air conditioning systems a minimum of once every two years.

❑ Keep filters clean or replace them regularly.

❑ Learn how to use your day/night thermostat for comfort and efficient energy use.

❑ If you have a zoned system (more than one furnace and separate controls) think through operating schedules and temperature settings to maximize comfort and minimize energy consumption.

❑ During cold days, open window coverings to allow the sun to warm your home. Close them when the sun begins to set.

❑ Limit use of your fireplace in extremely cold or windy weather when the chimney draft will draw room air out at an extreme rate.

❑ During the winter, humidifying the air in your home allows the air to retain more heat and is a general health benefit. Note: If condensation develops on your windows, you have taken a good thing too far and need to lower the setting on the humidifier. Avoid use of the humidifier when you are using your air conditioner.

❑ Ceiling fans cost little to operate and the moving air allows you to feel comfortable at temperatures several degrees higher.

❑ One hot days, close all windows and the window coverings on windows facing the sun to minimize solar heating and reduce demands on your air conditioner.

❑ Whole house fans draw cool outside air into the home through open windows, often effectively creating a comfortable temperature. Avoid running a whole house fan at the same time as air conditioning.

[Builder] Homeowner Manual

- Plan landscaping elements that support efficient energy use:

  - Deciduous trees provide shade during the summer and permit solar warming in winter.
  - Evergreen trees and shrubs can create a windbreak and reduce heating costs.
  - Position trees to shade the roof and still allow good air flow around the home.
  - Plant shrubs and trees to shade the air conditioner without obstructing air flow around the unit.

- Keep the garage overhead doors closed.

Water and Water Heater

- Set your water heater at 120 degrees if your dishwasher has a water booster heater. If not, set the water heater at 140 degrees.

- Follow the steps outlined in the manufacturer's directions for draining water from your water heater in order to remove accumulated hard-water scale that builds up inside the tank. Timing will depend on the nature of your water supply.

- Correct plumbing leaks, running toilets, or dripping faucets ASAP.

- Keep aerators clean.

- If you have a swimming pool, consider using solar heating power.

Appliances

- In selecting your home's appliances, compare the information on the (yellow and black) Energy Guide sticker. Sometimes spending a bit more up front can reduce operating costs over the life of the appliance, conserving energy at the same time.

- Use cold water when operating your disposal. This not only saves hot water you pay to heat, it preserves the disposal motor.

- When baking, preheat your oven just five minutes before you use it. When possible, bake several items at the same time or at least consecutively. Turn the oven off a few minutes before baking time is done.

- Microwave rather than using the range when possible, especially during hot weather.

- Run the dishwasher when it has a full load and use the air-dry cycle. Avoid regular use of the rinse and hold cycle.

[Builder] Homeowner Manual

❏ Turn electric burners off a few minutes before cooking is complete.

❏ Refrigerators with the freezer on top generally use significantly less energy than side-by-side models. Select an appropriate size for your needs; two small refrigerators use more energy than one large one.

Electrical

❏ Use compact fluorescent bulbs or fluorescent tubes where possible. Incandescent bulbs are the least efficient source of light.

❏ Turn lights and other electric items off when you finish using them or leave the room.

Maintenance

❏ Caulk in dry weather when temperatures are moderate. Check all locations, such as:

- Foundation penetrations (electrical, phone, water, cable tv, and gas line entrances)
- Around fans and vents
- Joints between door or window frames and siding

❏ Check weatherstripping on all exterior doors and adjust as needed. Ensure that door thresholds are a good fit—most are adjustable.

❏ After any activity in the attic, check that the insulation is evenly distributed.

Your Additional Reminders and Notes:

_____
_____
_____
_____
_____
_____

Appliance Service

This sheet is for your convenience. For warranty service on an appliance, contact the appropriate manufacturer directly at the service number provided in the appliance literature. You will need to supply the model and serial number (usually located on a small metal plate or seal attached to the appliance in an inconspicuous location), and the date of purchase (your closing date).

Closing Date _____

Appliance	*Manufacturer*	*Model #*	*Serial #*	*Service Phone #*
Range				
Range Hood				
Cooktop				
Oven				
Microwave				
Dishwasher				
Disposal				
Compacter				
Washer				
Dryer				
Refrigerator				
Freezer				

Home-Care Supplies

You will find that caring for your home is much easier if you have necessary tools and supplies on hand. As you review the maintenance information in this manual and in the manufacturer materials, note the materials and tools you will need. Note sizes, colors, brands, sources, and so on to create a convenient inventory that will make shopping for home-care products easier. You may wish to make copies of this form before filling it out.

For	Item	Color/Style	Size	Brand	Source	Notes

Maintenance Schedule

Begin care of your home with organized records, including information about all of its components and your furnishings. This information will make caring for you home easier, the records may be useful in completing tax returns, and will be valuable when you sell your home. Another worthwhile step is to inventory all equipment, appliances, furnishings, and personal belongings. A photo album containing pictures of each room is an excellent supplemental item. In addition to normal daily and weekly care, develop a schedule of preventative routines based on the information in this manual and the manufacturer literature you receive. A change of season creates special maintenance needs so plan for winterizing and summerizing your home.

Task/Notes	Frequency	J	F	M	A	M	J	J	A	S	O	N	D

Air Conditioning

Homeowner Use and Maintenance Guidelines

Air conditioning can greatly enhance the comfort of your home, but if it is used improperly or inefficiently, wasted energy and frustration will result. These hints and suggestions are provided to help you maximize your air conditioning system.

Your air conditioning system is a whole-house system. The air conditioning unit is the mechanism that produces cooler air. The air conditioning system involves everything inside your home including, for example, drapes, blinds, and windows.

Your home air conditioning is a closed system, which means that the interior air is continually recycled and cooled until the desired air temperature is reached. Warm outside air disrupts the system and makes cooling impossible. Therefore, you should keep all windows closed. The heat from the sun shining through windows with open drapes is intense enough to overcome the cooling effect of the air conditioning unit. For best results, close the drapes on these windows.

Time is very important in your expectations of an air conditioning system. Unlike a light bulb, which reacts instantly when you turn on a switch, the air conditioning unit only begins a process when you set the thermostat.

For example, if you come home at 6:00 p.m. when the temperature has reached 90 degrees F and set your thermostat to 75 degrees, the air conditioning unit will begin cooling, but will take much longer to reach the desired temperature. During the whole day, the sun has been heating not only the air in the house, but the walls, the carpet, and the furniture. At 6:00 p.m. the air conditioning unit starts cooling the air, but the walls, carpet, and furniture release heat and nullify this cooling. By the time the air conditioning unit has cooled the walls, carpet, and furniture, you may well have lost patience.

If evening cooling is your primary goal, set the thermostat at a moderate temperature in the morning while the house is cooler, allowing the system to maintain the cooler temperature. The temperature setting may then be lowered slightly when you arrive home, with better results. Once the system is operating, setting the thermostat at 60 degrees will *not* cool the home any faster and can result in the unit freezing up and not performing at all. Extended use under these conditions can damage the unit.

Adjust Vents

Maximize air flow to occupied parts of your home by adjusting the vents. Likewise, when the seasons change, readjust them for comfortable heating.

Compressor Level

Maintain the air conditioning compressor in a level position to prevent inefficient operation and damage to the equipment.

See also Grading and Drainage.

Humidifier

If a humidifier is installed on the furnace system, turn it off when you use the air conditioning; otherwise, the additional moisture can cause a freeze-up of the cooling system.

Manufacturer's Instructions

The manufacturer's manual specifies maintenance for the condenser. Review and follow these points carefully. Since the air conditioning system is combined with the heating system, follow the maintenance instructions for your furnace as part of maintaining your air conditioning system.

Temperature Variations

Temperatures may vary from room to room by several degrees Fahrenheit. This is due to such variables as floor plan, orientation of the home on the lot, type and use of window coverings, and traffic through the home.

Troubleshooting Tips: No Air Conditioning

Before calling for service, check to confirm that the:

- Thermostat is set to "cool" and the temperature is set below the room temperature.
- Blower panel cover is installed correctly for the furnace blower (fan) to operate. Similar to the way a clothes dryer door operates, this panel pushes in a button that lets the fan motor know it is safe to come on. If that button is not pushed in, the furnace will not operate.
- Air conditioner and furnace breakers on the main electrical panel are on. (Remember if a breaker trips you must turn it from the tripped position to the off position before you can turn it back on.)
- 220 switch on the outside wall near the air conditioner is on.
- Switch on the side of the furnace is on.
- Fuse in furnace is good. (See manufacturer literature for size and location.)
- Filter is clean to allow air flow.
- Vents in individual rooms are open.
- Air returns are unobstructed.
- Air conditioner has not frozen from overuse.

Even if the troubleshooting tips do not identify a solution, the information you gather will be useful to the service provider you call.

[Builder] Limited Warranty Guidelines

The air conditioning system should maintain a temperature of 78 degrees or a differential of 15 degrees from the outside temperature, measured in the center of each room at a height of 5 feet above the floor. Lower temperature settings are often possible, but neither the manufacturer nor [Builder] guarantee this.

Compressor

The air conditioning compressor must be in a level position to operate correctly. If it settles during the warranty period, [Builder] will correct this.

Coolant

The outside temperature must be 70 degrees F or higher for the contractor to add coolant to the system. If your home was completed during winter months, this charging of the system is unlikely to be complete and will need to be performed in the spring. Although we check and document this at orientation, your call to remind us is welcome in the spring.

Nonemergency

Lack of air conditioning service is not an emergency. Air conditioning contractors in our region respond to air conditioning service requests during normal business hours and in the order received.

Alarm System

Homeowner Use and Maintenance Guidelines

If your home selections included prewire for an alarm system, you will arrange for the final connection and activation after you move-in. The alarm company will demonstrate the system, instruct you in its use, and provide identification codes for your family. We recommend that you test the system each month.

[Builder] Limited Warranty Guidelines

[Builder] will correct wiring that does not perform as intended for the alarm system. [Builder] makes no representation that the alarm system will provide the protection for which it is installed or intended.

[Builder] Homeowner Manual

Appliances

Homeowner Use and Maintenance Guidelines

Please see page 8.5 and your *Appliance Service* information sheet.

[Builder] Limited Warranty Guidelines

We confirm that all appliance surfaces are in acceptable condition during your orientation. We assign all appliance warranties to you, effective on the date of closing. The appliance manufacturers warrant their products directly to you according to the terms and conditions of these written warranties.

Asphalt

Homeowner Use and Maintenance Guidelines

Asphalt is a flexible and specialized surface. Like any other surface in your home, it requires protection from things that can damage it. Over time, the effects of weather and earth movement will cause minor settling and cracking of asphalt. These are normal reactions to the elements and do not constitute improperly installed asphalt or defective material. Avoid using your driveway for one week after it is installed. Keep people, bicycles, lawn mowers, and other traffic off of it.

Chemical Spills

Asphalt is a petroleum product. Gasoline, oil, turpentine, and other solvents or petroleum products can dissolve or damage the surface. Wash such spills with soap and water immediately, and then rinse them thoroughly with plain water.

Hot Weather

Avoid any concentrated or prolonged loads on your asphalt, particularly in hot weather. High-heeled shoes, motorcycle or bicycle kickstands, trailers, or even cars left in the same spot for long periods can create depressions or punctures in asphalt.

Nonresidential Traffic

Prohibit commercial or other extremely heavy vehicles such as moving vans or other large delivery trucks from pulling onto your driveway. We design and install asphalt drives for conventional residential vehicle use only: family cars, vans, light trucks, bicycles, and so on.

Sealcoating

Exposure to sunlight and other weather conditions will fade your driveway, allowing the surface gravel material to be more visible. This is a normal condition and not a material or structural problem. You do not need to treat the surface of your asphalt driveway. However, if you choose to treat it, wait a minimum of 12 months and use a dilute asphalt emulsion, rather than the more common coal tar sealant. Hairline cracks will usually be filled by the sealing process. Larger cracks can be filled or patched with a sand and sealer mixture prior to resealing.

[Builder] Limited Warranty

We perform any asphalt repairs by overlay patching. [Builder] is not responsible for the inevitable differences in color between the patch and the original surface. Sealcoating can eliminate this cosmetic condition and is your responsibility.

Alligator Cracking

If cracking that resembles the skin of an alligator develops under normal residential use, [Builder] will repair it. If improper use, such as heavy truck traffic, has caused the condition, repairs will be your responsibility.

Settling

Settling next to your garage floor of up to 1.5 inches across the width of the driveway is normal. Settling or depressions elsewhere in the driveway of up to one inch in any 8-foot radius are considered normal. We will repair settling that exceeds these measurements.

Thermal Cracking

Your driveway will exhibit thermal cracking, usually during the first 12 months. These cracks help your driveway adapt to heating and freezing cycles. Cracks should be evaluated in the hottest months—July or August. We will repair cracks that exceed ½ inch in width.

Attic Access

Homeowner Use and Maintenance Guidelines

The attic space is neither designed nor intended for storage. We provide access to this area for maintenance of mechanical equipment that may traverse the attic space. When you perform needed tasks in the attic, use caution and avoid stepping off wood members onto the drywall. This can result in personal injury or damage to the ceiling below. Your limited warranty does not cover such injury or damage.

[Builder] Limited Warranty Guidelines

[Builder] and the local building department inspect the attic before your closing to confirm insulation is correct.

Brass Fixtures

Homeowner Use and Maintenance Guidelines

The manufacturer treats brass fixtures with a clear protective coating, electrostatically applied, to provide beauty and durability. This coating is not impervious to wear and tear. Atmospheric conditions, sunlight, caustic agents such as paints, and scratches from sharp objects can cause the protective coating to crack or peel, exposing the brass and resulting in spotting and discoloration.

Cleaning

Initial care of these products requires only periodic cleaning with a mild, nonabrasive soap and buffing with a soft cloth.

Corrosion

Unless you have ordered solid brass fixtures, the brass on your fixtures is a coating on top of a base metal. Water having a high mineral content is corrosive to any brass—coated or solid.

Polish

When peeling, spotting, or discoloration occurs, you can sometimes restore the beauty of the metal by completely removing the remaining coating and hand-polishing the item with a suitable brass polish. Applying a light coat of wax and buffing with a soft cloth helps maintain the gloss.

Tarnish

Like sterling silver, brass will gradually tarnish and eventually take on an antique appearance.

[Builder] Limited Warranty Guidelines

During the orientation we will confirm that brass fixtures are in acceptable condition. [Builder] does not warrant against corrosion damage to the external surfaces or internal workings of plumbing fixtures. This limitation includes solid brass or brass-coated fixtures.

Brick

Homeowner Use and Maintenance Guidelines

Brick is one of the most durable and lowest maintenance finishes for a home's exterior. A record of your brick color is included in your selection sheets.

Efflorescence

The white, powdery substance that sometimes accumulates on brick surfaces is called efflorescence. This is a natural phenomenon and cannot be prevented. In some cases, you can remove it by scrubbing with a stiff brush and vinegar. Consult your home center or hardware store for commercial products to remove efflorescence.

Tuck-Pointing

After several years, face brick may require tuck-pointing (repairing the mortar between the bricks). Otherwise, no regular maintenance is required.

Weep Holes

You may notice small holes in the mortar along the lower row of bricks. These holes allow moisture that has accumulated behind the brick to escape. Do not fill these weep holes or permit landscaping materials to cover them.

[Builder] Limited Warranty Guidelines

We check the brick-work during the orientation to confirm correct installation of designated materials.

Cracks

One time during the warranty period, we repair masonry cracks that exceed 3/16 inch.

Cabinets

Homeowner Use and Maintenance Guidelines

Your selection sheets are your record of the brand, style, and color of cabinets in your home. If you selected wood or wood veneer cabinets, expect differences in grain and color between and within the cabinet components due to natural variations in wood and the way it takes stain.

Cleaning

Products such as lemon oil or polishes that include scratch cover are suggested for wood cabinet care. Follow container directions. Use such products a maximum of once every 3 to 6 months to avoid excessive build-up. Avoid paraffin-based spray waxes and washing cabinets with water, as both will damage the luster of the finish.

Hinges

If hinges catch or drawer glides become sluggish, a small amount of silicone lubricant will improve their performance.

Moisture

Damage to cabinet surfaces and warping can result from operating appliances that generate large amounts of moisture (such as a crockpot) too near the cabinet. When operating such appliances, place them in a location that is not directly under a cabinet.

[Builder] Limited Warranty Guidelines

During the orientation, we will confirm that all cabinet parts are installed and that their surfaces are in acceptable condition.

Alignment

Doors, drawer fronts, and handles should be level and even.

Operation

Cabinets should operate properly under normal use.

Separations

We will correct gaps between cabinets and the ceiling or cabinets and the walls by caulking or other means if the gap exceeds 1/8 inch (locations behind appliances are excepted from this repair).

Warping

If doors or drawer fronts warp in excess of 1/4 inch within 24 inches, we will correct this by adjustment or replacement.

Wood Grain

Readily noticeable variations in wood grain and color are normal in all wood or wood veneer selections. Replacements are not made due to such variations.

Carpet

Homeowner Use and Maintenance Guidelines

Your selection sheets provide a record of the brand, style, and color of floor coverings in your home. Please retain this information for future reference. Refer to the various manufacturer's recommendations for additional information on the care of your floor coverings.

Burns

Take care of any kind of burn immediately. First snip off the darkened fibers. Then use a soapless cleaner and sponge with water. If the burn is extensive, talk with a professional about replacing the damaged area.

Cleaning

You can add years to the life of your carpet with regular care. Carpet wears out because of foot traffic and dirt particles that get trampled deep into the pile beyond the suction of the vacuum. The dirt particles wear down the fibers like sandpaper and dull the carpet. The most important thing you can do to protect your carpet is to vacuum it frequently.

Vacuum twice each week lightly and once a week thoroughly. Heavy traffic areas may require more frequent cleaning. A light vacuuming is three passes; a thorough job may need seven passes. A vacuum cleaner with a beater-bar agitates the pile and is more effective in bringing dirt to the surface for easy removal.

Vacuuming high-traffic areas daily helps keep them clean and maintains the upright position of the nap. Wipe spills and clean stains immediately. For best results, blot or dab any spill or stain; avoid rubbing. Test stain removers on an out-of-the-way area of the carpet, such as in a closet, to check for any undesirable effects.

Have your carpet professionally cleaned regularly, usually after 18 months in your home and then once a year after that.

Crushing

Furniture and traffic may crush a carpet's pile fibers. Frequent vacuuming in high-traffic areas and glides or cups under heavy pieces of furniture can help prevent this. Rotating your furniture to change the traffic pattern in a room promotes more even wear. Some carpets resist matting and crushing because of their level of fiber, but this does not imply or guarantee that no matting or crushing will occur. Heavy traffic areas such as halls and stairways are more susceptible to wear and crushing. This is considered normal wear.

Fading

Science has yet to develop a color that will not fade with time. All carpets will slowly lose some color due to natural and artificial forces in the environment. You can delay this process by frequently removing soil with vacuuming, regularly changing air filters in heating and air conditioning systems, keeping humidity and room temperature from getting too high, and reducing sunlight exposure with window coverings.

Filtration

If interior doors are kept closed while the air conditioning is operating, air circulation from the closed room flows through the small space at the bottom of the door. This forces the air over the carpet fibers, which in turn act as a filter, catching particulate pollution. Over time, a noticeable stain develops at the threshold.

See also Ghosting.

Fuzzing

In loop carpets, fibers may break. Simply clip the excess fibers. If it continues, call a professional.

Pilling

Pilling or small balls of fiber can appear on your carpet, depending on the type of carpet fiber and the type of traffic. If this occurs, clip off the pills. If they cover a large area, seek professional advice.

Rippling

With wall-to-wall carpeting, high humidity may cause rippling. If the carpet remains rippled after the humidity has left, have a professional restretch the carpeting using a power stretcher, not a knee-kicker.

Seams

Carpet usually comes in 12-foot widths, making seams necessary in most rooms. Visible seams are not a defect unless they have been improperly made or unless the material has a defect, making the seam appear more pronounced than normal. The more dense and uniform the carpet texture, the more visible the seams will be.

Carpet styles with low, tight naps result in the most visible seams. Seams are never more visible than when the carpet is first installed. Usually with time, use, and vacuuming the seams become less visible. You can see examples in the model homes of how carpet seams diminish after they have been vacuumed repeatedly and have experienced traffic.

Shading

Shading is an inherent quality of fine-cut pile carpets. Household traffic causes pile fibers to assume different angles; as a result, the carpet appears darker or lighter in these areas. A good vacuuming, which makes the pile all go in the same direction, provides a temporary remedy.

Shedding

New carpeting, especially pile, sheds bits of fiber for a period of time. Eventually these loose fibers are removed by vacuuming. Shedding usually occurs more with wool carpeting than with nylon or other synthetics.

Snags

Sharp-edged objects can grab or snag the carpet fiber. When this occurs, cut off the snag. If the snag is especially large, call a professional.

Sprouting

Occasionally you may find small tufts of fiber sprouting above carpet surface. Simply use scissors to cut off the sprout. Do not attempt to pull it, because other fibers will come out in the process.

Stains

No carpet is stain-proof. Although your carpet manufacturer designates your carpet as stain-resistant, some substances may still cause permanent staining. These include hair dyes, shoe polish, paints, and India ink. Some substances destroy or change the color of carpets, including bleaches, acne medications, drain cleaners, plant food, insecticides, and food or beverages with strongly colored natural dyes as found in some brands of mustard and herbal tea.

[Builder] Homeowner Manual

Refer to your care and maintenance brochures for recommended cleaning procedures for your particular fiber. Pretest any spot-removal solution in an inconspicuous area before using it in a large area. Apply several drops of the solution, hold a white tissue on the area, and count to ten. Examine both tissue and carpet for dye transfer and check for carpet damage.

Static

Cooler temperatures outside often contribute to static electricity inside. To avoid the problem, look for carpets made with anti-static. You can also install a humidifier to help control static build-up.

[Builder] Limited Warranty Guidelines

During your orientation, we will confirm that your carpet is in acceptable condition. We will correct stains or spots noted at this time by cleaning, patching, or replacement. [Builder] is not responsible for dye lot variations if replacements are made.

Edges

Edges of carpet along moldings and edges of stairs should be held firmly in place. In some areas, metal or other edging material may be used where carpet meets another floor covering.

Seams

Carpet seams will be visible. [Builder] will repair any gaps or fraying.

Caulking

Homeowner Use and Maintenance Guidelines

Time and weather will shrink and dry caulking so that it no longer provides a good seal. As routine maintenance, check the caulking and make needed repairs. Caulking compounds and dispenser guns are available at hardware stores. Read the manufacturer's instructions carefully to be certain that you select an appropriate caulk for the intended purpose.

Colored Caulk

Colored caulking is available where larger selections are provided. As with any colored material, dye lots can vary.

Latex Caulk

Latex caulking is appropriate for an area that requires painting, such as along the stair stringer or where wood trim meets the wall.

Silicone Caulk

Caulking that contains silicone will not accept paint; it works best where water is present, for example, where tub meets tile or a sink meets a countertop.

[Builder] Limited Warranty Guidelines

During the orientation we confirm that appropriate areas are adequately caulked.

One-Time Repair

We will touch up caulking one time during your materials and workmanship period. We suggest that this be performed with your 11-month service.

See also Countertops, Expansion and Contraction, Stairs, and Wood Trim.

Ceramic Tile

Homeowner Use and Maintenance Guidelines

Your selection sheets include the brand and color of your ceramic tile.

Cleaning

Ceramic tile is one of the easiest floor coverings to maintain. Simply vacuum when needed. Occasionally, a wet mopping with warm water may be appropriate. Avoid adding detergent to the water. If you feel a cleaning agent is required, use a mild solution of warm water and dishwasher crystals (they will not result in a heavy, difficult-to-remove lather on the grout). Rinse thoroughly.

The ceramic tile installed on walls or countertops in your home may be washed with any nonabrasive soap, detergent, or tile cleaner. Abrasive cleaners will dull the finish.

Grout Discoloration

Clean grout that becomes yellowed or stained with a fiber brush, cleanser, and water. Grout cleansers and whiteners are available at most hardware stores.

Sealing Grout

Sealing grout is your decision and responsibility. Once grout has been sealed, ongoing maintenance of that seal is necessary and limited warranty coverage on grout that has been sealed is void.

Separations

Expect slight separations to occur in the grout between tiles. This grout is for decorative purposes only; it does not hold the tile in place. Cracks in the grout can be filled using premixed grout purchased from flooring or hardware stores. Follow package directions.

Tile around bathtubs or countertops may appear to be pulling up after a time. This is caused by normal shrinkage of grout or caulk and shrinkage of wood members as they dry out. If this occurs, the best remedy is to purchase tub caulk or premixed grout from a hardware store. Follow directions on the container. This maintenance is important to protect the underlying surface from water damage.

[Builder] Limited Warranty Guidelines

During the orientation we confirm that tile and grout areas are in acceptable condition. We will repair or replace cracked, badly chipped, or loose tiles noted at that time. [Builder] is not responsible for variations in color or discontinued patterns. New grout may vary in color from the original.

One-Time Repair

Cracks appearing in grouting of ceramic tiles at joints or junctions with other materials are commonly due to shrinkage. [Builder] will repair grouting, if necessary, one time during the first year. We are not responsible for color variations in grout or discontinued colored grout. Any grouting or caulking that is needed after that time is your responsibility.

Concrete Flatwork

Homeowner Use and Maintenance Guidelines

By maintaining good drainage, you protect your home's foundation and the concrete flatwork: the basement floor, porch, patio, driveway, garage floor, and sidewalks.

Concrete slabs are floating—they are not attached to the home's foundation walls. These are not a structural (load-bearing) element of the home and are covered by the one year material and workmanship warranty.

We install a flexible collar around the top of the furnace plenum. Gas and water lines include flexible connections, and drain lines have slip joints. The basement stairs do not rest on the floor and the support posts under the I-beam are separated from the floor slab. [Builder] incorporates all of these details in the construction of the basement floor because we know the floor will move in response to the soils. Movement of the basement slab or any concrete slab results in cracking. Minimize this movement by following [Builder's] landscaping recommendations, the objective of which is to prevent moisture from reaching soils around and under the home.

Cleaning

Avoid washing exterior concrete slabs with cold water from an outside faucet when temperatures are high and the sun has been shining on the concrete. The abrupt change in temperature can damage the surface bond of the concrete. We recommend sweeping for keeping exterior concrete clean. If washing is necessary, do this when temperatures are moderate. Repeated cleaning of the garage floor by hosing can increase soil movement by allowing water to penetrate any existing cracks. We recommend sweeping to clean the garage floor.

Cracks

A concrete slab 10 feet across shrinks approximately 5/8 inch as it cures. Some of this shrinkage shows up as cracks. Cracking of concrete flatwork also results from temperature changes that cause expansion and contraction.

During the summer, moisture finds its way under the concrete along the edges or through cracks in the surface. In winter, this moisture forms frost that can lift the concrete, increasing the cracking. Maintaining drainage away from all concrete slabs will minimize cracking from this cause.

As cracks occur, seal them with a waterproof concrete caulk (available at hardware or home improvement stores) to prevent moisture from penetrating to the soil beneath.

Expansion Joints

We install expansion joints to help control expansion. However, as the concrete shrinks during the curing process, moisture can penetrate under the concrete and lift the expansion joint. When this occurs, fill the resulting gap with a gray silicone sealant, which you can purchase at most hardware stores.

Heavy Vehicles

Prohibit commercial or other extremely heavy vehicles such as moving vans and other large delivery trucks from pulling onto your driveway. We design and install concrete drives for conventional residential vehicle use only: family cars, vans, light trucks, bicycles, and so on.

Ice, Snow, and Chemicals

Driving or parking on snow creates ice on the drive, which magnifies the effects of snow on the concrete surface. Remove ice and snow from concrete slabs as promptly as possible after snow storms. Protect concrete from abuse by chemical agents such as pet urine, fertilizers, radiator overflow, repeated hosing, or de-icing agents, such as road salt that can drip from vehicles. All of these items can cause spalling (chipping of the surface) of concrete.

Post-Tension Slabs

If your home is built on a post-tension slab, avoid any action that penetrates the concrete. The risk of hitting a cable or tendon, which is under considerable tension, makes such actions dangerous.

Sealer

A concrete sealer, available at paint stores, will help you keep an unpainted concrete floor clean. Do not use soap on unpainted concrete. Instead, use plain water and washing soda or, if necessary, a scouring powder.

[Builder] Limited Warranty Guidelines

Concrete slabs are floating—they are not attached to the home's foundation walls. Because these slabs are not a structural (load-bearing) element of the home, they are excluded from coverage under the structural warranty. The limited warranty coverage is for one year unless the requirements of your loan state otherwise.

Color

Concrete slabs vary in color. [Builder] provides no correction for this condition.

Cracks

If concrete cracks reach 3/16 of an inch in width or vertical displacement, [Builder] will patch or repair them one time during the warranty year. Subsequently, concrete slab maintenance is your responsibility. If you prefer to have the slab replaced, we will obtain a price for you and assist in scheduling the work upon receipt of your payment. However, we advise against this expense since the new slab will crack as well.

Finished Floors

[Builder] will correct cracks, settling, or heaving that rupture finish floor materials that we installed as part of the home as you originally purchased it.

Level Floors

Concrete floors in the habitable areas of the home will be level to within 1/4 inch within any 32-inch measurement with the exception of an area specifically designed to slope toward a floor drain.

Separation

[Builder] will correct separation of concrete slabs from the home if separation exceeds one inch.

Settling or Heaving

[Builder] will repair slabs that settle or heave in excess of 2 inches or if such movement results in negative drainage (toward the house) or hazardous vertical displacement.

Spalling (Surface Chips)

Causes of spalling include repeated hosing of concrete for cleaning, animal urine, radiator overflow, fertilizer, uncleared snow and ice, ice-melting agents, and road salts from vehicles. Repair of spalling is a home maintenance task.

Standing Water

Water may stand on exterior concrete slabs for several hours after precipitation or from roof run-off. [Builder] will correct conditions that cause water to remain longer than 12 hours unless it is from roof run-off of melting snow or ice.

Condensation

Homeowner Use and Maintenance Guidelines

When warm, moist air comes into contact with cooler surfaces, the moisture condenses. Outside we see this as dew; inside you may see it as a layer of moisture on glass windows and doors. This condensation comes from high humidity within the home combined with low outside temperatures and inadequate ventilation. Family lifestyle significantly influences two out of three of these conditions.

Humidifier Operation

If your home includes a humidifier, closely observe manufacturer's directions for its use. Instructions to turn the humidifier off during air conditioning season are typical. Moderate settings in winter can maintain desired comfort levels without contributing too much moisture to your home. You may need to experiment to find the correct level for your family's lifestyle.

New Construction

Some experts have estimated that a typical new home contains 50 gallons of water. Water is part of lumber, concrete, drywall texture, paint, caulk, and other materials used in building. Wet weather during construction adds more. This moisture evaporates into the air as you live in your home–adding to the moisture generated by normal living activities. Over time, this source of moisture will diminish.

Normal Activities

As you live in your home, your daily lifestyle contributes to the moisture in the air also. Cooking, laundry, baths and showers, aquariums, plants, and so on all add water to the air in your home. Likewise, your daily routine can mitigate the amount of moisture in your home and reduce condensation on interior surfaces.

Temperature

Avoid setting your thermostat at extreme temperatures. Heating your home will cause the materials to dry out faster, generating more moisture into the air; drying the materials out too fast also increases shrinkage cracks and separations.

Ventilation

Develop the habit of using exhaust fans in bathrooms and over the stove. When weather conditions permit, open windows so fresh air can circulate through your home. Keep the dryer exhaust hose clean and securely connected.

See also Ventilation.

[Builder] Limited Warranty Guidelines

Condensation results from weather conditions and a family's lifestyle. [Builder] has no control over these factors. The limited warranty coverage excludes condensation.

Countertops

Homeowner Use and Maintenance Guidelines

Use a cutting board to protect your counters when you cut or chop. Protect the counter from heat and from extremely hot pans. If you cannot put your hand on it, do not put it on the counter. Do not use countertops as ironing boards and do not set lighted cigarettes on the edge of the counter.

Caulking

The caulking between the countertop and the wall, along the joint at the backsplash (the section of counter that extends a few inches up the wall along the counter area), and around the sink may shrink, leaving a slight gap. Maintaining a good seal in these locations is important to keep moisture from reaching the wood under the laminates and to prevent warping.

Cleaning

Avoid abrasive cleaners that will damage the luster of the surface.

Mats

Rubber drain mats can trap moisture beneath them, causing the laminated plastic to warp and blister. Dry the surface as needed.

Wax

Wax is not necessary, but it can be used to make counters gleam.

See also Ceramic Tile.

[Builder] Limited Warranty Guidelines

During your orientation we confirm that all countertops are in acceptable condition. We repair noticeable surface damage such as chips, cracks, and scratches noted on the orientation list. Repair of surface damage noted subsequent to this is one of your home maintenance responsibilities.

Laminates

Laminated countertops will have one or more discernible seams. [Builder] will repair gaps or differential at the seams that exceed 1/16 inch.

Manufactured Marble

Edges should be smooth and even. Where backsplash joints occur at corners, the top edges should be even within 1/16 inch.

Separation from Wall

Separation of countertops from walls, backsplash, and around sinks results from normal shrinkage of materials. [Builder] will recaulk these areas one time during the materials and workmanship warranty. Subsequently caulking will be your home maintenance responsibility.

Crawl Space

Homeowner Use and Maintenance Guidelines

The crawl space is not intended as a storage area for items that could be damaged by moisture. Wood stored in a crawl space can attract termites.

You may notice slight dampness in the crawl space. Landscaping that is correctly installed helps prevent excessive amounts of water from entering crawl spaces. Report standing water to [Builder] for inspection.

See also Ventilation.

[Builder] Limited Warranty Guidelines

During the orientation we will check the condition of soils in the crawl space. Soils in the crawl space may be damp but should not have standing water. Provided that you have not altered the drainage nor caused excessive moisture to accumulate and remain in this area with incorrect landscaping, [Builder] will correct the conditions that result in persistent standing water.

Dampproofing

Homeowner Use and Maintenance Guidelines

We spray your foundation walls with an asphalt waterproofing material. Although we make every effort to assure a dry basement, during times of excessive moisture, you may notice some dampness. Over time, natural compaction of soils in the backfill areas will usually eliminate this condition. Careful maintenance of positive drainage will also protect your basement from this condition.

[Builder] Limited Warranty Guidelines

[Builder] will correct conditions that allow actual water to enter the basement unless the cause is improper installation of landscaping or failure to adequately maintain drainage.

Decks

Homeowner Use and Maintenance Guidelines

Wood decks add to the style and function of your home and are a high maintenance part of your home's exterior.

Effects of Exposure

Wood decks are subject to shrinkage, cracking, splitting, cupping, and twisting. Nails or screws may work lose and will need routine maintenance. Plan to inspect your decks regularly, a minimum of once each year, and provide needed attention promptly to maintain an attractive appearance and forestall costly repairs. [Builder] recommends that you treat or restain your decks annually to keep them looking their best.

Foot Traffic

As you use your decks, abrasives and grit on shoes can scratch or dent the wood surface. Regular sweeping and mats can mitigate this but will not completely prevent it.

Outdoor Furniture

The surface of the decking can be damaged by moving grills, furniture, or other items. Use caution when moving such items to prevent scratches, gouges, and so on.

Sealing or Water Repellent

To prolong the life and beauty of your deck, treat it periodically with a water repellent or wood preservative. Local home centers or hardware stores offer several products to consider for this purpose. Always follow manufacturer directions carefully.

Snow and Ice

Heavy snow or ice that remains on the deck over long periods increases wear and tear on the deck. Prompt removal can reduce adverse effects. Use caution in shoveling to avoid needless scratching of the deck boards.

Stain

Exposed wood decks have been stained with a semi-transparent oil stain to protect and beautify the wood. Each board takes the same stain differently and variations in color will be readily noticeable. Over time, with exposure to weather and use, further variations in color will occur.

[Builder] Limited Warranty Guidelines

Exposed wood decks are constructed to meet structural and functional design. During the orientation, we will confirm that the wood decks are in satisfactory condition.

Color Variation

Color variations are a natural result of the way in which wood accepts stain and are excluded from limited warranty coverage.

Replacement Boards or Rails

Shrinkage, cracking, splitting, cupping, and twisting are natural occurrences in wood decks and are excluded from limited warranty coverage. In extreme situations where personal safety is involved, if [Builder] provides replacement of boards or rails, the new material will not match existing pieces that have been exposed to elements and use. [Builder] does not provide corrections when problems occur due to lack of normal maintenance.

Doors and Locks

Homeowner Use and Maintenance Guidelines

The doors installed in your home are wood products subject to such natural characteristics of wood as shrinkage and warpage. Natural fluctuations caused by humidity and the use of forced air furnaces, showers, and dishwashers, interior doors may occasionally require minor adjustments.

Bifold Doors

Interior bifolds sometimes stick or warp because of weather conditions. Apply a silicone lubricant to the tracks to minimize this inconvenience.

Exterior Finish

To ensure longer life for your exterior wood doors, plan to refinish them at least once a year. Stained exterior doors with clear finishes tend to weather faster than painted doors. Treat the finish with a wood preserver every three months to preserve the varnish and prevent the door from drying and cracking. Reseal stained exterior doors whenever the finish begins cracking or crazing.

Failure to Latch

If a door will not latch because of minor settling, you can correct this by making a new opening in the jamb for the latch plate (remortising) and raising or lowering the plate accordingly.

Hinges

You can remedy a squeaky door hinge by removing the hinge pin and applying a silicone lubricant to it. Avoid using oil, as it can gum up or attract dirt. Graphite works well as a lubricant but can create a gray smudge on the door or floor covering beneath the hinge if too much is applied.

Keys

Keep a duplicate privacy lock key where children cannot reach it in the event a youngster locks him- or herself in a room. The top edge of the door casing is often used as a place to keep the key. A small screwdriver or similarly shaped device can open some types of privacy locks.

Locks

Lubricate door locks with graphite or other waterproof lubricant. Avoid oil, as it will gum up.

Shrinkage

Use putty, filler, or latex caulk to fill any minor separations that develop at mitered joints in door trim. Follow with painting. Panels of wood doors shrink and expand in response to changes in temperature and humidity. Touching up the paint or stain on unfinished exposed areas is your home maintenance responsibility.

Slamming

Slamming doors can damage both doors and jambs and can even cause cracking in walls. Teach children not to hang on the doorknob and swing back and forth; this works loose the hardware and causes the door to sag.

Sticking

The most common cause of a sticking door is the natural expansion of lumber caused by changes in humidity. When sticking is caused by swelling during a damp season, do not plane the door unless it continues to stick after the weather changes.

Before planing a door because of sticking, try two other steps: first, apply either a paste wax, light coat of paraffin, or candle wax to the sticking surface; or second, tighten the screws that hold the door jamb or door frame. If planing is necessary even after these measures, use sandpaper to smooth the door and paint the sanded area to seal against moisture.

Warping

If a door warps slightly, keeping it closed as much as possible often returns it to normal.

Weather Stripping

Weather stripping and exterior door thresholds occasionally require adjustment or replacement.

[Builder] Limited Warranty Guidelines

During the orientation we confirm that all doors are in acceptable condition and correctly adjusted. [Builder] will repair construction damage to doors noted on the orientation list.

Adjustments

Because of normal settling of the home, doors may require adjustment for proper fit. [Builder] will make such adjustments.

Panel Shrinkage

Panels of wood doors shrink and expand in response to changes in temperature and humidity. Although touching up the paint or stain on unfinished exposed areas is your home maintenance responsibility, [Builder] will repair split panels that allow light to be visible.

Warping

[Builder] will repair doors that warp in excess of 1/4 inch.

Drywall

Homeowner Use and Maintenance Guidelines

Slight cracking, nail pops, or seams may become visible in walls and ceilings. These are caused by the shrinkage of the wood and normal deflection of rafters to which the drywall is attached.

Ceilings

The ceilings in your home are easy to maintain: periodically remove dust or cobwebs as part of your normal cleaning and repaint as needed.

Repairs

With the exception of the one-time repair service provided by [Builder], care of drywall is your maintenance responsibility. Most drywall repairs can be easily made. This work is best done when you redecorate the room.

Repair hairline cracks with a coat of paint. You can repair slightly larger cracks with spackle or caulk. To correct a nail pop, reset the nail with a hammer and punch. Cover it with spackle, which is available at paint and hardware stores. Apply two or three thin coats. When dry, sand the surface with fine-grain sandpaper, and then paint. You can fill indentations caused by sharp objects in the same manner.

[Builder] Limited Warranty Guidelines

During the orientation, we confirm that drywall surfaces are in acceptable condition.

One Time Repairs

One time during the materials and workmanship warranty, [Builder] will repair drywall shrinkage cracks and nail pops and will touch up the repaired area using the same paint color that was on the surface when the home was delivered. Touch-ups will be visible.

Repainting the entire wall or the entire room to correct this is your choice and responsibility. You are also responsible for custom paint colors or wallpaper that has been applied subsequent to closing. Due to the effects of time on paint and wallpaper, as well as possible dye lot variations, touch-ups are unlikely to match the surrounding area.

Lighting Conditions

[Builder] does not repair drywall flaws that are only visible under particular lighting conditions.

Related Warranty Repairs

If a drywall repair is needed as a result of poor workmanship (such as blisters in tape) or other warranty-based repair (such as a plumbing leak), [Builder] completes the repair by touching up the repaired area with the same paint that was on the surface when the home was delivered. If more than one-third of the wall is involved, we will repaint the wall corner to corner. You are responsible for custom paint colors or wallpaper that has been applied subsequent to closing. The effects of time on paint and wallpaper, as well as possible dye lot variations, mean touch-up may not match the surrounding area.

Easements

Homeowner Use and Care Guidelines

Easements are areas where such things as utility supply lines can pass through your property. They permit service to your lot and adjacent lots, now and in the future. Your lot will also include drainage easements, meaning the runoff from adjacent lots passes across your property. Likewise, water form your property may run across a neighboring lot. Easements are recorded and are permanent.

Trees, shrubs, gardens, play equipment, storage sheds, fences or other items which you install in or across these easements may be disturbed if service entities–such as the gas, electric, or phone companies–need access to lines for repairs or to connect service to nearby homesites.

Utility companies, the United States Postal Service, and others have the right to install equipment in easements. These might include streetlights, mailboxes, or junction boxes to name a few. Neither [Builder] nor you as the homeowner have the authority to prevent, interfere with, or alter these installations. Plans for the location of such items are subject to change by the various entities involved. Because they have no obligation to keep [Builder] informed of such changes, we are unable to predict specific sites that will include such equipment.

See also Property Boundaries.

Electrical System

Homeowner Use and Maintenance Guidelines

Know the location of the breaker panel; it includes a main shut-off that controls all the electrical power to the home. Individual breakers control the separate circuits. Each breaker is marked to help you identify which breaker is connected to which major appliances, outlets, or other service. Should a failure occur in any part of your home, always check the breakers in the main panel box.

Breakers

Circuit breakers have three positions: on, off, and tripped. When a circuit breaker trips, it must first be turned off before it can be turned on. Switching the breaker directly from tripped to on will not restore service.

Breakers Tripping

Breakers trip because of overloads caused by plugging too many appliances into the circuit, a worn cord or defective appliance, or operating an appliance with too high a voltage requirement for the circuit. The starting of an electric motor can also trip a breaker.

If any circuit trips repeatedly, unplug all items connected to it and reset. If it trips when nothing is connected to it, you need an electrician. If the circuit remains on, one of the items you unplugged is defective and will require repair or replacement.

Buzzing

Fluorescent fixtures use transformer action to operate. This action sometimes causes a buzzing.

Fixture Location

We install light fixtures in the locations indicated on the plans. Moving fixtures to accommodate specific furniture arrangements or room use is your responsibility.

GFCI (Ground-Fault Circuit-Interrupters)

GFCI receptacles have a built-in element that senses fluctuations in power. Quite simply, the GFCI is a circuit breaker. Building codes require installation of these receptacles in bathrooms, the kitchen, outside, and the garage (areas where an individual can come into contact with water while holding an electric appliance or tool). Heavy appliances such as freezers or power tools will trip the GFCI breaker.

Caution: Never plug a refrigerator or food freezer into a GFCI-controlled outlet. The likelihood of the contents being ruined is high and the limited warranty does not cover such damage.

Each GFCI circuit has a test and reset button. Once each month, press the test button. This will trip the circuit. To return service, press the reset button. If a GFCI breaker trips during normal use, it may indicate a faulty appliance and you will need to investigate the problem. One GFCI breaker can control up to three or four outlets.

Grounded System

Your electrical system is a three-wire grounded system. Never remove the bare wire that connects to the box or device.

Light Bulbs

You are responsible for replacing burned-out bulbs other than those noted during your orientation.

Luminous Light Panels

Translucent panels covering ceiling lights are made of polystyrene plastic. To clean, gently push up, tilting the panel slightly and remove it from the fixture frame. Wash with a diluted (1 to 2 percent) solution of mild detergent and warm water. Do not rinse; the soap film that remains reduces static electricity that attracts dust.

Over time, the plastic panel may yellow and will become brittle and may need to be replaced if it cracks or breaks. Replacement material can be found at home center and hardware stores. Most suppliers will cut the panel to fit so if you need to purchase a replacement, be sure to note the size you need.

Bulbs for these fixtures can be purchased at home centers or hardware stores. Avoid exceeding the wattage indicated inside the fixture.

Modifications

If you wish to make any modifications, contact the electrician listed on the Emergency Phone Numbers you receive at the orientation. Having another electrician modify your electrical system during the warranty period can void that portion of your limited warranty.

Outlets

If an outlet is not working, check first to see if it is controlled by a wall switch or GFCI. Next, check the breaker.

If there are small children in the home, install safety plugs to cover unused outlets. This also minimizes the air infiltration that sometimes occurs with these outlets. Teach children to never touch electrical outlets, sockets, or fixtures.

[Builder] Homeowner Manual

Underground Cables

Before digging, check the location of buried service leads by calling the local utility locating service. In most cases, wires run in a straight line from the service panel to the nearest public utility pad. Maintain positive drainage around the foundation to protect electrical service connections.

Under- or Over-Cabinet Lights

The selection of optional under- or over-cabinet lighting provides either task lighting or atmosphere to your kitchen. We suggest you note the size and type of bulbs in these fixtures and keep replacements on hand.

TROUBLESHOOTING TIPS: NO ELECTRICAL SERVICE

No Electrical Service Anywhere in the Home

Before calling for service, check to confirm that the:

- Service is not out in the entire area. If so, contact the utility company.
- Main breaker and individual breakers are all in the on position.

No Electrical to One or More Outlets

Before calling for service, check to confirm that the

- Main breaker and individual breakers are all in the on position.
- Applicable wall switch is on
- GFCI is set (see details on GFCIs, earlier in this section)
- Item you want to use is plugged in
- Item you want to use works in other outlets
- Bulb in the lamp is good

Even if the troubleshooting tips do not identify a solution, the information you gather will be useful to the service provider you call.

[Builder] Limited Warranty Guidelines

During the orientation, we confirm that light fixtures are in acceptable condition and that all bulbs are working. [Builder]'s limited warranty excludes any fixture you supplied.

Designed Load

[Builder] will repair any electrical wiring that fails to carry its designed load to meet specifications. If electrical outlets, switches, or fixtures do not function as intended, [Builder] will repair or replace them.

GFCI (Ground-Fault Circuit-Interrupters)

[Builder] is not responsible for food spoilage that results from your plugging refrigerators or freezers into a GFCI outlet.

Power Surge

Power surges are the result of local conditions beyond the control of [Builder] and are excluded from limited warranty coverage. These can result in burned-out bulbs or damage to sensitive electronic equipment such as TVs, alarm systems, and computers. Damage resulting from lightning strikes are excluded from limited warranty coverage.

Evaporative Cooler

Homeowner Use and Maintenance Guidelines

An evaporative cooler is an efficient way to cool your home in a dry climate. Cooling occurs as air is pulled across wet pads allowing the system to circulate cool moist air into your home. Read and follow the manufacturer directions for use and care.

Distribution Lines

About twice a year, check the connections and distribution lines for obstructions or leaks.

Drain Reservoir

On a monthly basis, drain the water from the reservoir and replace it.

Pad Replacement

At least once each year, replace the pads with new ones.

[Builder] Limited Warranty Guidelines

Proper performance of the evaporative cooler will be confirmed at your orientation. Refer to the manufacturer's limited warranty for complete information regarding warranty coverage on your evaporative cooler.

Expansion and Contraction

Homeowner Use and Maintenance Guidelines

Changes in temperature and humidity cause all building materials to expand and contract. Dissimilar materials expand or contract at different rates. This movement results in separation between materials, particularly dissimilar ones. You will see the effects in small cracks in drywall and in paint, especially where moldings meet drywall, at mitered corners, and where tile grout meets tub or sink. While this can alarm an uninformed homeowner, it is normal.

Shrinkage of the wood members of your home is inevitable and occurs in every new home. Although this is most noticeable during the first year, it may continue beyond that time. In most cases, caulk and paint are all that you need to conceal this minor evidence of a natural phenomenon. Even though properly installed, caulking shrinks and cracks. Maintenance of caulking is your responsibility.

[Builder] Limited Warranty

[Builder] provides one-time repairs to many of the effects of expansion and contraction. See individual categories such as drywall and caulk for details.

Fencing

Homeowner Use and Maintenance Guidelines

Depending on the community in which your home is located, fencing may be included with your home, it may be an optional item, or it may be an item you consider adding after your move-in. When [Builder] installs fencing as part of your new home, we confirm its good condition during your orientation. All types of fencing require some routine attention.

Drainage

In planning, installing, and maintaining fencing, allow existing drainage patterns to function unimpeded. When installing a fence, use caution in distributing soil removed to set posts to avoid blocking drainage swales. Plan enough space under the bottom of a wood fence for water to pass through.

Homeowner Association Design Review

If you choose to add fencing after moving into your new home, keep in mind the need to obtain approval form the Design Review Committee of your homeowners association. Specific requirements about style, height, position on the lot are described in the current design review guidelines which you can obtain from a committee member. Special requirements apply to homes on corner lots where drivers must have adequate visibility. Additionally, in some communities, zoning laws may impact private fencing. Your responsibilities include checking on such details.

[Builder] recommends that you engage the services of professionals to install your fence. Be certain to inform a fence installer of all design review requirements.

See also Property Boundaries.

Variation

Height and location of [Builder] installed fences will vary with lot size, topography, and shape. [Builder] must meet the requirements of the Design Review process just as any homeowner would.

Wood Fences

The lumber used to construct wood fences is rough cedar. Over time it will crack, warp, and split. Unless extreme, these conditions require no action on your part. As the wood ages and shrinks, nails may come loose and require attention. Also check the posts and any gates twice a year and tighten hardware or make needed adjustments.

Wrought Iron Fencing

Wrought iron is subject to rusting, if it is not maintained. Use touch-up paint on any scratches or chips. Inspect the fence twice a year and touch-up as needed, then plan to repaint the entire fence every one to two years to keep it looking its best.

As with wood fencing, prevent sprinklers from spraying your wrought iron fence or rails. Check monthly to confirm that water does not stand around the fence posts. Make corrections to drainage as needed to prevent this.

[Builder] Limited Warranty

If fencing is part of your home purchase, we will confirm the acceptable condition of the fence during your orientation. [Builder] will correct fence posts that become lose during the warranty period. Be aware that damage to fencing caused by severe weather should be referred to your homeowner insurance company and is specifically excluded form warranty coverage.

Fireplace

Homeowner Use and Maintenance Guidelines

See also Fire Prevention.

Most of us feel a fireplace is an excellent way to create a warm, cozy atmosphere. However, without sufficient information, your use of the fireplace can result in heat (and dollars) being wasted. To help prevent that, consider the following points.

Look upon burning a fire as a luxury that adds much to the atmosphere but just a little to the heat in a home. About 10 percent of the heat produced by a fire is radiated into the house. In many older homes, the air used by the fireplace for combustion is replaced with cold outside air drawn in through cracks around doors and windows. However, your home is constructed so tightly that this does not happen. We install a fresh air vent to supply the fireplace with combustion air and reduce the amount of heated air the fire draws from your house. Open this vent before starting the fire as you do the damper.

Close the damper and cold air vent when the fireplace is not in use. Leaving these open is equivalent to having an open window in the house. If the fire is still burning, but you are finished enjoying it, use glass doors to prevent heated air from being drawn up the chimney until your damper can be closed.

One caution on the use of glass doors: do not close them over a roaring fire, especially if you are burning hard woods (such as oak or hickory) because this could break the glass. Also, when closing the doors over a burning fire, open the mesh screens first. This prevents excessive heat build-up on the mesh, which might result in warping or discoloration.

Your objective in building a fire should be a clean, steady, slow-burning fire. Begin with a small fire to allow the components of the fireplace to heat up slowly. Failure to do so may damage the fireplace and can void the warranty. Start the fire by burning kindling and newspaper under the grate; two to three layers of logs stacked with air space between, largest logs to the rear, works best. One sheet of paper burned on top of the stack will help the chimney start to draw. Any logs 6 inches in diameter or larger should be split.

Caution: Do not burn trash in the fireplace and never use any type of liquid fire starter.

Remove old ashes and coals from under the grate when completely cool. A light layer is desirable as an insulator and will help to reflect heat.

Chimney Cleaning

Creosote and other wood-burning by-products accumulate inside the flue over a period of time. This build-up can be a fire hazard. The way you use your fireplace and the type of wood you burn determine the frequency of your chimney cleanings. For instance, burning soft woods or improperly seasoned woods necessitates more frequent cleaning. Hire a qualified chimney sweep for this cleaning.

Spark Arrester

If the spark arrester becomes clogged, the diminished air flow will affect the performance of the fireplace and may be a fire hazard. Have the arrester cleaned professionally when needed.

Gas Fireplace

[Builder] offers direct-vent gas fireplaces. If you ordered this type of fireplace, it is demonstrated during the orientation. Read and follow all manufacturer's directions.

A slight delay between turning the switch on and flame ignition is normal. The flames should ignite gently and silently. If you notice any deviation from this and any gas smell, immediately shut off the switch and report it to the gas company.

Excessive winds can cause a downdraft, which can blow out the pilot, requiring you to relight it before using the fireplace.

Caution: The exterior vent cover for a direct-vent gas fireplace becomes extremely hot when the fireplace is operating.

[Builder] Limited Warranty Guidelines

Fireplaces are not intended to be the sole heat source in the home. The fireplace should function properly when [Builder]'s and the manufacturer's directions are followed.

Chimney Separation

Separation of a brick chimney from a newly constructed home may occur. [Builder] will repair separation from the main structure in excess of ½ inch in 10 feet. Caulking is acceptable in most cases.

Cracks

Normal shrinkage of mortar results in hairline cracks in masonry. [Builder] will repair cracks that exceed 1/8 inch in width. The repair consists of pointing or patching and the mortar color will be matched as closely as possible, but expect some variation.

Exterior masonry may have chips, irregular surfaces, and color variations, which occur during manufacturing, shipping, or handling. Unless such conditions affect the structural integrity of the home, no repair is provided.

Discoloration

Discoloration of the firebox or brick is a normal result of use and requires no corrective action. Mortar-style fireplaces may develop cracks due to temperature changes and other factors.

Downdraft

Although extremely high winds can result in a downdraft, this condition should be temporary and occasional. We will determine and correct continuous malfunction if caused by a construction or design defect.

Glass Doors

During the orientation we confirm that glass fireplace doors, when included with the home, are in acceptable condition.

Water Infiltration

In unusually heavy or prolonged precipitation, especially when accompanied by high winds, some water can enter the home through the chimney. The limited warranty excludes this occurrence.

Foundation

Homeowner Use and Maintenance Guidelines

We install the foundation of your home according to the recommendations of our consulting engineer. The walls of the foundation are poured concrete with steel reinforcing rods. To protect your home's foundation, follow guidelines for installation and maintenance of landscaping and drainage in this manual.

Cracks

Even though an engineer designed the foundation and we constructed it according to engineering requirements, surface cracks can still develop in the wall. Surface cracks are not detrimental to the structural integrity of your home. If a crack develops in a foundation wall that allows water to come through, follow the procedures for submitting a warranty claim.

Dampness

Due to the amount of water in concrete, basements may be damp. Condensation can form on water lines and drip onto the floor.

Future Construction in Basement

If you decide to perform additional construction in the basement, obtain guidelines from a licensed engineer, obtain a building permit, and comply with all codes and safety requirements. [Builder] does not warrant that you will be able to obtain such a permit because of the possibility that building codes may change.

[Builder] Limited Warranty Guidelines

The foundation of your home has been designed and installed according to the recommendations of an engineer. The walls of the foundation are poured concrete with steel reinforcing rods.

Cosmetic Imperfections

Slight cosmetic imperfections in foundation walls, such as a visible seam where two pours meet or slight honeycombing (aggregate visible), are possible and require no repair unless they permit water to enter.

Cracks

Shrinkage or backfill cracks are not unusual in foundation walls, especially at the corners of basement windows. [Builder] will seal cracks that exceed 1/8 inch in width.

Leaks

[Builder] will correct conditions that permit water to enter the basement, provided you have complied with the drainage, landscaping, and maintenance guidelines.

Garage Overhead Door

Homeowner Use and Maintenance Guidelines

Since the garage door is a large, moving object, periodic maintenance is necessary.

Light Visible

Garage overhead doors cannot be airtight. Some light will be visible around the edges and across the top of the door. Weather conditions may result in some precipitation entering around the door as well as some dust especially until most homes in the community have landscaping installed.

Lock

If the lock becomes stiff, apply a silicone or graphite lubricant. Do not use oil on a lock, as it will stiffen in winter and make the lock difficult to operate.

Lubrication

Every 6 months, apply a lubricant such as silicone spray to all moving parts: track, rollers, hinges, pulleys, and springs. Avoid over lubricating to prevent drips on vehicles or the concrete floor. At the same time, check to see that all hardware is tight and operating as intended without binding or scraping.

Opener

To prevent damage to a garage door opener, be sure the door is completely unlocked and the rope-pull has been removed before using the operator. If you have an opener installed after closing on your home, we suggest that you order it from the company that provided and installed the garage door to assure uninterrupted warranty coverage. Be familiar with the steps for manual operation of the door in the event of a power failure.

If [Builder] installed a door opener as one of your selections, during orientation we demonstrate the electric eye that provides a safety stop in the event someone crosses through the opening while the overhead door is in motion. Use care not to place tools or other stored items where they interfere with the function of the electric eye.

Expect to replace the battery in the garage opener remote controls about once a year. The battery is usually a 9 volt.

Painting

Repaint the garage door when you repaint your home, or more often if needed to maintain a satisfactory appearance.

Safety

Follow the manufacturer's instructions for safe and reliable operation. Do not allow anyone except the operator near the door when it is in motion. Keep hands and fingers away from all parts of the door except the handle. Do not allow children to play with or around the door.

For your safety, after the expiration of the one-year warranty, have any needed adjustments made by a qualified specialist. The door springs are under a considerable amount of tension and require special tools and knowledge for accurate and safe servicing. Have the door inspected by a professional garage door technician after any significant impact to the door.

Sag

The garage door may sag slightly due to its weight and span. This will stabilize after the panels have dried.

[Builder] Limited Warranty Guidelines

The garage door should operate smoothly and with reasonable ease. The door can become misaligned and require adjustment, which [Builder] will provide unless the problem is caused by the installation of a garage door opener subsequent to closing on the home.

Gas Shut-Offs

Homeowner Use and Maintenance Guidelines

You will find shut-offs on gas lines near their connection to each item that operates on gas. In addition, there is a main shut-off at the meter. We point these out during the orientation.

Gas Leak

If you suspect a gas leak, leave the home and call the gas company immediately for emergency service.

[Builder] Limited Warranty Guidelines

The gas company is responsible for leaks up to the meter. [Builder] will correct leaks from the meter into the home.

Ghosting

Homeowner Use and Maintenance Guidelines

Recent feedback from homeowners (in both old and new homes) regarding black sooty stains which develop on surfaces in homes (on carpet, walls, ceilings, appliances, mirrors, and around area rugs—to list a few examples) have caused much investigation and research.

The conclusion of the research and laboratory tests has been that the majority of this staining or "ghosting" results from pollution of the air in the home caused by burning scented candles. Incomplete combustion of hydrocarbons as these candles burn contributes a considerable amount of soot to the air. This sooty substance then settles or accumulates on surfaces of the home. The sooty deposits are extremely difficult to remove; on some surfaces (light-colored carpet, for instance), they are impossible to clean completely away.

The popularity of scented candles has increased many-fold in recent years. If this is an activity that is part of your lifestyle, we caution you about the potential damage to your home. When this condition results from homeowners burning candles or other lifestyle choices, the resulting damage is excluded from our limited warranty coverage.

See also Carpet/Filtration.

Grading and Drainage

Homeowner Use and Maintenance Guidelines

The final grades around your home have been inspected and approved for proper drainage of your lot. Our surveyor completes a drainage certification and then the local building authorities as well as [Builder] inspect the site. Yards drain from one to another. You and your neighbors share an overall drainage plan for the community. Use caution when installing landscaping, fencing, or additions t your home to prevent causing water problems on adjacent lots.

Drainage

Typically, the grade around your home should slope 1 foot in the first 10 feet, tapering to a 2 percent slope. In most cases, drainage swales do not follow property boundaries. Maintain the slopes around your home to permit the water to drain away from the home as rapidly as possible. This is essential to protect your foundation. Failure to do so can result in major structural damage and will void your warranty.

Exterior Finish Materials

Maintain soil levels 6 inches below siding, stucco, brick, or other exterior finish materials. Contact with the soil can cause deterioration of the exterior finish material and encourages pest infestations.

Roof Water

Do not remove the splash blocks or downspout extensions from under the downspouts. Keep these in place at all times, sloped so the water drains away from your home quickly.

Rototilling

Rototilling can significantly change drainage swales. You can minimize this by rototilling parallel to the swales rather than across them.

Settling

The area we excavated for your home's foundation was larger than the home to allow room to work. In addition, some trenching is necessary for installation of utility lines. Although we replaced and compacted the soil, it does not return to its original density. Some settling will occur, especially after prolonged or heavy rainfall or melting of large amounts of snow. Settling can continue for several years. Inspect the perimeter of your home regularly for signs of settling and fill settled areas as needed to maintain positive drainage.

Subsurface Drains

Occasionally [Builder] installs a subsurface drain to ensure that surface water drains from a yard adequately. Keep this area and especially the drain cover clear of debris so that the drain can function as intended.

See also Landscaping.

[Builder] Limited Warranty Guidelines

We established the final grade to ensure adequate drainage away from the home. Maintaining this drainage is your responsibility. If you alter the drainage pattern after closing, or if changes in drainage occur due to lack of maintenance, the limited warranty is void.

Backfill Settlement

Backfilled or excavated areas around the foundation and at utility trenches should not interfere with the drainage away from your home. If these areas settle during the first year, [Builder] will fill the areas one time and subsequently will provide you with fill dirt to maintain positive drainage.

Erosion

[Builder] is not responsible for weather-caused damage to unlandscaped yards after the final grade has been established or the closing date, whichever occurs last.

New Sod

New sod installation and the extra watering that accompanies it can cause temporary drainage problems, as can unusually severe weather conditions.

Recommendations

[Builder] documents the grades that exist at the time of delivery of your home or as soon thereafter as possible. The ground must be dry and free of frost to make these determinations. Once final grades are set, [Builder] will inspect drainage problems reported in writing during the warranty period, compare grades to those originally established, and advise you on corrective actions you might take.

Soil Information

We provide soil information when the purchase agreement is written or as soon thereafter as it becomes available. Landscaping recommendations are designed based on soils and engineering reports and thus may vary slightly.

[Builder] Homeowner Manual

Swales

[Builder] does not alter drainage patterns to suit individual landscape plans. Typically, a lot receives water from and passes water on to other lots, so changes in grade often affect adjacent or nearby lots. [Builder] advises against making such changes. After heavy rain or snow, water may stand in swales up to 48 hours.

Under Concrete

[Builder] will fill visible sunken areas under concrete during the first year.

Winter Grading

Due to weather conditions, especially during winter and early spring, the final grade may not have been established at the time of closing. We document the status of your grading at the time of delivery. When conditions permit, grading work will continue. Confirm that we have completed your grading before beginning landscaping.

Gutters and Downspouts

Homeowner Use and Maintenance Guidelines

Check gutters periodically and remove leaves or other debris. Materials that accumulate in gutters can slow water drainage from the roof, cause overflows, and clog the downspouts.

Extensions or Splashblocks

Extensions should discharge outside of rock or bark beds so that water is not dammed behind the edging materials that might be used.

Ladders

Use caution when leaning ladders against gutters, as this may cause dents.

Leaks

If a joint between sections of gutter drips, caulk the inside joint using a commercial gutter caulking compound available at hardware stores.

Paint

Gutters and downspouts are painted to match your home. You should repaint them when you repaint your home.

Snow and Ice

Clear excess snow from downspouts as soon as possible to allow the gutter to drain and to prevent damage. Severe ice or snow build-up can damage gutters, and such damage is not covered by the limited warranty.

See also Roof/Ice Dam.

[Builder] Limited Warranty Guidelines

Gutters over 3 feet long are installed with a slight slope so that roof water will flow to the downspouts.

Leaks

We correct leaks that occur during the warranty period.

Overflow

Gutters may overflow during periods of excessively heavy rain. This is expected and requires no repair.

Standing Water

Small amounts of water (up to one inch) will stand for short periods of time in gutters immediately after rain. No correction is required for these conditions.

Hardware

Homeowner Use and Maintenance Guidelines

Doorknobs and locks should operate correctly with little attention. Over time, they may need slight adjustments due to normal shrinkage of the framing. Occasionally, you may need to tighten screws or lubrication.

[Builder] Limited Warranty Guidelines

We confirm that all hardware is in acceptable condition during orientation. The limited warranty excludes repairs for cosmetic damage subsequent to the orientation.

[Builder] will repair hardware items that do not function as intended.

[Builder] Homeowner Manual

Hardwood Floors

Homeowner Use and Maintenance Guidelines

In daily care of hardwood floor, preventive maintenance is the primary goal.

Cleaning

Sweep on a daily basis or as needed. Never wet-mop a hardwood floor. Excessive water causes wood to expand and can possibly damage the floor. When polyurethane finishes become soiled, damp-mop with a mixture of 1 cup vinegar to one gallon of warm water. When damp-mopping, remove all excess water from the mop. Check with the hardwood company if your floor has a water-based finish.

Dimples

Placing heavy furniture or dropping heavy or sharp objects on hardwood floors can result in dimples.

Filmy Appearance

A white, filmy appearance can result from moisture, often from wet shoes or boots.

Furniture Legs

Install proper floor protectors on furniture placed on hardwood floors. Protectors will allow chairs to move easily over the floor without scuffing. Regularly clean the protectors to remove any grit that may have accumulated.

Humidity

Wood floors respond noticeably to changes in humidity in your home. Especially during winter months the individual planks or pieces expand and contract as water content changes. A humidifier helps but does not eliminate this reaction.

Mats and Area Rugs

Use protective mats at the exterior doors to help prevent sand and grit from getting on the floor. Gritty sand is wood flooring's worst enemy. However, be aware that rubber backing on area rugs or mats can cause yellowing and warping of the floor surface.

Recoat

If your floors have a polyurethane finish, you may want to have an extra coat of polyurethane applied by a qualified contractor within six months to one year. The exact timing will depend on your particular lifestyle. If another finish was used, refer to the manufacturer's recommendations.

Separation

Expect some shrinkage around heat vents or any heat-producing appliances, or during seasonal weather changes.

See also Warping.

Shoes

Keep high heels in good repair. Heels that have lost their protective cap (thus exposing the fastening nail) will exert over 8,000 pounds of pressure per square inch on the floor. That's enough to damage hardened concrete; it will mark your wood floor.

Spills

Clean up food spills immediately with a dry cloth. Use a vinegar-and-warm-water solution for tough food spills.

Splinters

When floors are new, small splinters of wood can appear.

Sun Exposure

Exposure to direct sunlight can cause irreparable damage to hardwood floors. To preserve the beauty of your hardwood floors, install and use window coverings in these areas.

Traffic Paths

A dulling of the finish in heavy traffic areas is likely.

Warping

Warping will occur if the floor repeatedly becomes wet or is thoroughly soaked even once. Slight warping in the area of heat vents or heat-producing appliances is also typical.

[Builder] Homeowner Manual

Wax

Waxing and the use of products like oil soap are neither necessary nor recommended. Once you wax a polyurethane finish floor, recoating is difficult because the new finish will not bond to the wax. The preferred maintenance is preventive cleaning and recoating annually or as needed to maintain the desired luster.

[Builder] Limited Warranty Guidelines

During the orientation we will confirm that hardwood floors are in acceptable condition. We will correct any readily noticeable cosmetic defects listed during the orientation. You are responsible for routine maintenance of hardwood floors.

Separations

Shrinkage will result in separations between the members of hardwood floors. If these exceed 1/8 inch, [Builder] will fill them one time. [Builder] is not responsible for removing excess filler that appears on the surface if the boards expand due to subsequent changes in humidity and expel the filler.

Heating System: Gas Forced Air

Homeowner Use and Maintenance Guidelines

Good maintenance of your furnace can save energy dollars and prolong the life of the furnace. Carefully read and follow the manufacturer's literature on use and maintenance. The guidelines here include general information only.

Adjust Vents

Experiment with the adjustable registers in your home to establish the best heat flow for your lifestyle. Generally, you can reduce the heat in seldom-used or interior rooms. This is an individual matter and you will need to balance the system for your own family's needs.

Avoid Overheating

Do not overheat your new home. Overheating can cause excessive shrinkage of framing lumber and may materially damage the home. In the beginning, use as little heat as possible and increase it gradually.

Blower Panel (Fan cover)

You need to position the blower panel cover correctly for the furnace blower (fan) to operate. This panel compresses a button that tells the blower it is safe to operate. Similar to the way a clothes dryer door operates, this panel pushes in a button that lets the fan motor know it is safe to come on. If that button is not pushed in, the furnace will not operate.

Combustion Air

Furnaces we install in basements or in utility closets over crawl spaces include a combustion air duct. The outside end of this duct is covered with a screen to minimize insect or animal from entering the duct. Cold air coming in though this duct means it is functioning as it should.

Caution: Never cover or block the combustion air vent in any way. Outside air is needed to supply the furnace with sufficient oxygen. Blocking the combustion air vent will cause the furnace to draw air down the vent pipe and pull poisonous gases back into your home.

Duct Cleaning

Exercise caution before spending money on professional ductwork cleaning services. A study by the EPA found no proof that ductwork cleaning improves indoor air quality, nor was evidence found that it prevents health problems. For more information contact the EPA and request document EPA-402-K-97-002. Or you can view this information on their Website: www.epa.gov/iaq/pubs/.

Ductwork Noise

Some popping or pinging sounds are the natural result of ductwork heating and cooling in response to airflow as the system operates.

Filter

A clean filter will help to keep your home clean and reduce dusting chores. Remember to change or clean the filter monthly during the heating season (year-round if you also have air conditioning). A clogged filter can slow airflow and cause cold spots in your home. Although it takes less than one minute to change the filter, this is one of the most frequently overlooked details of normal furnace care.

Buy filters in large quantity for the sake of convenience. You will find the size and type printed along the edge of the filter that in your furnace.

If you have a permanent, washable, removable filter, you need to clean this monthly. Use water only to clean the filter, tap to dry or air dry, and leave unit off for a brief period. Do not use soaps or detergents on the filter.

[Builder] Homeowner Manual

Furnished Home

The heating system was designed with a furnished home in mind. If you move in during the cooler part of the year and have not yet acquired all of your draperies and furnishings, the home may seem cooler than you would expect.

Fuse

Some furnaces have a fuse directly above the on-off switch. This fuse is an S10, S12, or S15 fuse. It absorbs any spikes in the line such as close electrical strikes or power surges. Unlike old fuses that burn out and clearly indicate that they are blown, these fuses, similar to automobile fuses, have a spring that depresses when tripped. Unless you have examined these quite carefully before, it may be hard to determine if the fuse has blown. We suggest that you buy some extra fuses of the same size to have on hand.

Gas Odor

If you smell gas, call the gas company immediately.

Odor

A new heating system may emit an odor for a few moments when you first turn it on. An established system may emit an odor after being unused for an extended time (such as after the summer months if you do not use air conditioning). This is caused by dust that has settled in the ducts and should pass quickly.

On-Off Switch

The furnace has an on-off blower switch. This switch looks like a regular light switch and is located in a metal box outside the furnace. When turned off, this switch overrides all furnace commands and shuts down the blower. This is usually done only when maintenance service is being performed, although young children have been known to turn the furnace off using this switch. (If your furnace is a high-efficiency model, it does not have a pilot or an on-off switch.)

Pilot

On models with manually lit pilots, lighting the furnace pilot involves several steps. First, remove the cover panel to expose the pilot. Then rotate the on-off-pilot knob to pilot. When the knob is in this position, you can depress the red button.

While depressing the red button, hold a match at the pilot. Once the pilot lights, continue to hold the red button down for 30 to 60 seconds. When you release the red button, the pilot should stay lit. If it does not, wait several minutes to allow any gas to dissipate from the furnace area and repeat the entire process. If the pilot stays lit, rotate the on-off pilot knob to the on position. Reinstall the cover panel. You can find these instructions on a sticker on the furnace and in the manufacturer's literature.

Registers

Heat register covers are removable and adjustable. You are responsible for adjusting the dampers in these covers to regulate the heat flow within the home. Registers in the rooms farther away from the furnace will usually need to be opened wider.

Return Air Vents

For maximum comfort and efficient energy use, arrange furniture and draperies to allow unobstructed airflow from registers and to cold air returns.

Temperature

Depending on the style of home, temperatures can normally vary from floor to floor as much as 10 degrees or more on extremely cold days. The furnace blower will typically cycle on and off more frequently and for shorter periods during severe cold spells.

Thermostat

The furnace will come on automatically when the temperature at the thermostat registers below the setting you have selected. Once the furnace is on, setting the thermostat to a higher temperature will not heat the home faster. Thermostats are calibrated to within plus or minus 5 degrees.

Trial Run

Have a trial run early in the fall to test the furnace. (The same applies to air conditioning in the spring.) If service is needed, it is much better to discover that before the heating season.

TROUBLESHOOTING TIPS: NO HEAT

Before calling for service, check to confirm that the:

- Thermostat is set to "heat" and the temperature is set above the room temperature.
- Blower panel cover is installed correctly for the furnace blower (fan) to operate. This panel compresses a button that tells the blower it is safe to operate. Similar to the way a clothes dryer door operates, this panel pushes in a button that lets the fan motor know it is safe to come on. If that button is not pushed in, the furnace will not operate.

[Builder] Homeowner Manual

- Breaker on the main electrical panel is on. (Remember, if a breaker trips you must turn it from the tripped position to the off position before you can turn it back on.)
- Switch on the side of the furnace is on.
- Fuse in furnace is good. (See manufacturer literature for size and location.)
- Gas line is open at the main meter and at the side of the furnace.
- Filter is clean to allow airflow.
- Vents in individual rooms are open.
- Air returns are unobstructed.

Even if the troubleshooting tips do not identify a solution, the information you gather will be useful to the service provider you call.

[Builder] Limited Warranty Guidelines

We will install heating systems according to local building codes, as well as to engineering designs of the particular model home.

Adequacy of the system is determined by its ability to establish a temperature of 70 degrees F, as measured in the center of the room, 5 feet above the floor. In extremely cold temperatures (10 degrees below or colder), the system should be able to maintain a temperature differential of 80 degrees from the outside temperature.

Duct Placement

The exact placement of heat ducts may vary from those positions shown in similar floor plans.

Ductwork

Although the heat system is not a sealed system, the ductwork should remain attached and securely fastened. If it becomes unattached, [Builder] will repair as needed.

Furnace Sounds

Expansion or contraction of metal ductwork results in ticking or popping sounds. While eliminating all these sounds is impossible, [Builder] will correct oil canning. (Oil canning occurs when a large area of sheet metal like those found in air ducts makes a loud noise as it moves up and down in response to temperature changes.)

Thermostat

Thermostats are calibrated to plus or minus 5 degrees.

[Builder] Homeowner Manual

Heating System: Heat Pump

Homeowner Care and Maintenance

If your home contains a heat pump system, you should be aware of the performance characteristics unique to these systems. As with any system, read the manufacturer's literature and follow all instructions for efficient operation and maintenance of your system. Clean or replace filters once a month. Provide professional service for your system at least once every two years.

Air Circulation Across Coils

Keep the outside unit clear of any materials that would interfere with air circulation. Snow, ice, landscaping materials, trash, leaves, and other accumulating items can cause inefficiency or damage the unit.

Air Conditioning and Heating

A heat pump system operates differently from a gas forced-air furnace. The same system provides both heat and air conditioning. This is possible because a refrigerant flows back and forth in the coils of the heat pump, controlled by a reversing valve. In the heating mode, the heat pump removes heat from the outside air and transfers it to the inside air. In the cooling mode, it does just the opposite, removing heat from the inside air and discharging it outside of the home. The thermostat inside your home controls this heating or cooling activity.

Air Temperature at Vents

Do not expect dramatic temperature differences in the air coming from the vents as is common with other kinds of systems. The coils used in a heat pump system operate at lower temperatures than those common in a gas forced-air system. As a result, for example, in the heat mode, air from the supply vents will typically range from 85 to 90 degrees F. The vents will not feel hot, though the air discharged is warmer than the air in the room by as much as 20 degrees.

Auxiliary Heat System

At lower outside temperatures, less heat is available for the heat pump to draw from the exterior air. Therefore, from time to time the auxiliary heat system will come on to maintain the temperature you set at the thermostat. The auxiliary system will also come on whenever the temperature at the thermostat is moved 1.5 degrees or more at one time. If the light stays on when the outside temperature is more than 30 degrees F, contact a service person.

[Builder] Homeowner Manual

Defrost Cycle

When the heat pump is operating in the heat mode, the coils outside may reach below freezing temperatures. Moisture in the air will condense into frost and accumulate on the coils under these circumstances. From time to time, the system will go into defrost mode to clear accumulated frost from the coils. This is a normal part of the operation of the system and will occur automatically.

During the defrost cycle, the outside fan will stop temporarily. The temperature of airflow into the home will be a bit lower during the defrost cycle. The defrost cycle can only occur once every 90 minutes and lasts no longer than 10 minutes.

Night Setback

Unless you have a night setback thermostat designed to work with a heat pump system, do not turn the thermostat down in the evenings. Adjust the temperature a fraction of a degree at a time until a comfortable, permanent setting is found.

Register Adjustment

Registers will require adjustment from time to time to maximize your family's comfort. Do not completely close off more than one supply register at a time. This can restrict the airflow too much and reduce the efficiency of the system. A good technique is to completely open all the vents, then gradually move the temperature setting up until the coolest room is comfortable. Once the coolest room is comfortable, gradually close the vents in the warmer rooms until all rooms are comfortable as well. Reverse the process for air conditioning.

Return Air Vents

As with any heating system, return air vents must be clear so the air flows through the ducts unimpeded. Avoid placing furniture where it blocks the return air vents.

TROUBLESHOOTING TIPS: NO HEAT OR AUXILIARY HEAT STAYS ON WHEN OUTSIDE TEMPERATURE IS 30 DEGREES OR ABOVE

Before calling for service, check to confirm that the:

- Thermostat is set to "heat" and the temperature is set above the room temperature.
- Breaker on the main electrical panel is on. (Remember, if a breaker trips you must turn it from the tripped position to the off position before you can turn it back on.)
- Filter is clean to allow airflow.
- Vents in individual rooms are open.
- Air returns are unobstructed.
- Outside unit is not blocked by snow or other materials.
- Outside coil does not have an excessive ice build-up.

[Builder] Homeowner Manual

Even if the troubleshooting tips do not identify a solution, the information you gather will be useful to the service provider you call.

[Builder] Limited Warranty Guidelines

Refer to the manufacturer's limited warranty for information regarding warranty coverage.

Humidifier

Homeowner Use and Maintenance Guidelines

Operate a humidifier only with the furnace, not with the air conditioner. If you notice condensation on windows, the humidifier should be adjusted to a lower setting. Clean the moisture pad according to the manufacturer's instructions and suggested timetable.

[Builder] Limited Warranty Guidelines

Refer to the manufacturer's limited warranty for information regarding coverage of the humidifier.

Insulation

Homeowner Use and Maintenance Guidelines

The effectiveness of blown insulation is diminished if it is uneven. As the last step in any work done in your attic (for example, the installation of speaker wire), you should confirm that the insulation lays smooth and even. Do not step on drywall ceilings, because this can result in personal injury or damage to the drywall.

Electrical outlets normally emit noticeable amounts of cold air when outside temperatures are low.

[Builder] Limited Warranty Guidelines

[Builder] will install insulation to meet or exceed the building codes applicable at the time of construction and outlined as part of your purchase agreement.

Landscaping

Homeowner Use and Maintenance Guidelines

Providing complete details on landscape design is beyond the scope of this manual. Many excellent books, videos, and computer software programs are available that offer your this information. Local nurseries and landscape professionals can also assist you.

In planning your landscaping, think of proportion, texture, color, mature size, maintenance needs, soft and hard surfaces, lighting, fencing, edging, and water requirements. A beautiful yard requires considerable planning and regular attention. Most homeowners take years to achieve the yard they want. Planning to install items in stages can spread the cost and work over several seasons.

Whatever the source of your design, plan to install the basic components of your landscaping as soon after closing as weather permits. In addition to meeting your homeowner association requirements to landscape in a timely manner, well-designed landscaping prevents erosion and protects the foundation of your home.

Additions

Before installing patio additions or other permanent improvements, consider soil conditions in the design and engineering of your addition.

Backfill

We construct the foundation of your home beginning with an excavation into the earth. When the foundation walls are complete, the area surrounding them is backfilled. Soil in this area is not as compact as undisturbed ground. Water can penetrate through the backfill area to the lower areas of your foundation. This can cause potentially severe problems such as wet basements, cracks in foundation walls, and floor slab movement. Avoid this through proper installation of landscaping and good maintenance of drainage.

Backfill areas will settle and require prompt attention to avoid damage to your home and voiding of the structural warranty.

Keep downspout extensions in the down position to channel roof runoff away from the foundation area of your home. Routine inspection of downspouts, backfill areas, and other drainage components is an excellent maintenance habit.

See also Grading and Drainage.

Bark or Rock Beds

Do not allow edging around decorative rock or bark beds to dam the free flow of water away from the home. You can use a nonwoven landscape fabric between the soil and rock or bark to restrict weed growth while still permitting normal evaporation of ground moisture.

Erosion

Until your yard is established and stable, erosion will be a potential concern. Heavy rains or roof runoff can erode soil. The sooner you restore the grade to its original condition, the less damage will occur.

Erosion is of special concern in drainage swales. If swales become filled with soil runoff, they may not drain the rest of the yard, causing further problems. Correcting erosion is your responsibility. You may need to protect newly planted seed with erosion matting or reseed to establish grass in swales. It can take several years to fully establish your lawn in such challenging areas.

First 5 Feet

Place no plants of any type or sprinkler heads within 5 feet of your home.

Hired Contractors

You are responsible for changes to the drainage pattern made by any landscape, concrete, deck, or pool contractor. Discuss drainage with any company you hire to do an installation in your yard. Do not permit them to tie into existing drainage pipes without approval from [Builder].

Natural Areas

During construction, we remove construction debris from natural areas. Removing dead wood, tree limbs, fallen trees, or other natural items is your responsibility.

Planning

Locate plants and irrigation heads out of the way of pedestrian or bicycle traffic and car bumpers. Space groves of trees or single trees to allow for efficient mowing and growth. Group plants with similar water, sun, and space requirements together.

Plant Selection

Plant with regard to your local climate. Favor native over exotic species. Consider ultimate size, shape, and growth of the species.

See also Property Lines.

[Builder] Homeowner Manual

Requirements

Check with your local building department and homeowners association before designing, installing, or changing landscaping for any regulations that they require you to follow.

Seeded Lawns

If lawn seeding is part of your home purchase, consider this just the first step in establishing your yard. Remember that the forces of nature are far stronger than grass seed. You will need to overseed at some point, perhaps more than once. Heavy storms can cause washouts and erosion that you will need to correct. It generally takes at least three growing seasons to establish a good lawn, longer if weather conditions are difficult or if you do not have the time to devote to lawn care.

Before over-seeding, remember to fill any slight depressions with a light layer of topsoil. Minimize traffic of all kinds on newly seeded areas and avoid weed killer for at least 120 days. Keep the seed moist, not wet.

Sod

Newly placed sod requires extra water for several weeks. Water in the cool part of the day (ideally just before sunrise) at regular intervals for the first three weeks. Be aware that new sod and the extra watering it requires can sometimes create drainage concerns (in your yard or your neighbor's) that will disappear when the yard is established and requires normal watering.

Soil Mix

Provide good soil mixes with sufficient organic material. Use mulch at least 3 inches deep to hold soil moisture and to help prevent weeds and soil compaction.

In areas with high clay content, prepare the soil before installing your grass. First cover the soil with 2 inches of sand and 1 inch of manure that is treated and odorless. Rototill this into the soil to a depth of 6 inches (rototill parallel to the swales). Whether you use seed or sod, this preparation helps your lawn to retain moisture and require less water. Installing a lawn over hard soil permits water to run off with little or no penetration and your lawn will derive minimal benefit from watering or rain.

Apply appropriate fertilizer and weed and pest controls as needed for optimal growth. Investigate organic compounds for additional protection of the environment.

Sprinkler System

If [Builder] included a sprinkler system with your home, we will arrange to have the installer demonstrate the system and make final adjustments shortly after you move in. The installer will note and correct any deficiencies in the system at the same time. Whether we install your sprinkler or you install it yourself, keep these points in mind.

You are responsible for routine cleaning and adjusting of sprinkler heads as well as shutting the system down in the fall. Failure to drain the system before freezing temperatures occur can result in broken lines, which will be your responsibility to repair.

Conduct weekly operational checks to ensure proper performance of the system. Direct sprinkler heads away from the home. Trickler- or bubbler-type irrigation systems are not recommended for use adjacent to your home.

Automatic timers permit you to water at optimum times whether you are at home, away, awake, or asleep. The amount of water provided to each zone can be accurately and consistently controlled and easily adjusted with a timed system. Check the system after a power outage and keep a battery in place if your system offers that as a backup.

Stones

The soil in your area may have stones and rocks. Removing these naturally occurring elements is a maintenance activity. If [Builder] installs seed or sod, large rocks will be picked up and surface raking performed. You will need to provide continued attention to this condition as you care for your yard.

Trees

 [Builder] values trees as one of the features that make up an attractive community and add value to the homes we build. We take steps to protect and preserve existing trees in the area of your home. In spite of our efforts, existing trees located on construction sites can suffer damage from construction activities, which manifest months after the completion of construction.

Damage to existing trees can be caused by such things as compaction of soil in the root zone, changing patterns of water flow on the lot, disturbing the root system, and removing other trees to make room for the home. The newly exposed tree may react to conditions it is unaccustomed to. Caring for existing trees, including pruning dead branches or removing these trees altogether is your responsibility.

Remember to water trees during the summer or during warm dry periods in the winter.

[Builder] Homeowner Manual

Mulch around trees and avoid tilling or planting flower beds around trees. This is especially important while trees are recovering form the construction process.

Trees and other plant materials that exist on the lot when construction begins and are not part of any landscaping installed by [Builder] are excluded from warranty coverage.

Utility Lines

A slight depression may develop in the front lawn along the line of the utility trench. To correct this, roll back the sod, spread topsoil underneath to level the area, and then relay the sod.

Before any significant digging, check the location of buried service leads by calling the local utility locating service. In most cases, wires and pipes run in a straight line from the main service to the pubic supply.

See also Easements.

Waiting to Landscape

If you leave ground unlandscaped, it erodes. Correcting erosion that occurs after closing is your responsibility.

Weeds

Weed swill appear in your new lawn whether seed or sod is used. Left unlandscaped, your yard will quickly begin to show weeds. When soil is disturbed, dormant seeds come to the surface and germinate. The best control is a healthy lawn, achieved through regular care and attention.

Xeriscape®

[Builder] recommends careful consideration of landscape design and selection of planting materials to minimize the demands of your yard on water supplies. Detailed information about Xeriscape® is available from reputable nurseries. This has the triple benefit of helping the environment, saving on water bills, and reducing the amount of moisture that can reach your foundation.

[Builder] Limited Warranty

Landscape materials we install are warranted for one growing season. We will confirm the healthy condition of all plant materials during the orientation. Maintaining landscaping is your responsibility.

[Builder] Homeowner Manual

Mildew

Homeowner Use and Maintenance Guidelines

Mildew is a fungus that spreads through the air in microscopic spores. They love moisture and feed on surfaces or dirt. On siding, they look like a layer of dirt. To determine whether you are dealing with mildew or dirt, wipe the surface with a cloth or sponge dampened with bleach. If the bleach causes the surface to lose its dark appearance, you are most likely seeing mildew.

Cleaning mildew from your home is your responsibility. Solutions that remove mildew are available from local paint or home improvement stores. Wear protective eyewear and rubber gloves for this task; the chemicals that remove mildew are unfriendly to humans.

[Builder] Limited Warranty Guidelines

We will remove any mildew noted during the orientation. [Builder] warranty excludes mildew.

Mirrors

Homeowner Use and Maintenance Guidelines

To clean your mirrors, use any reliable liquid glass cleaner or polisher available at most hardware or grocery stores. Avoid acidic cleaners and splashing water under the mirror; either can cause the silvering to deteriorate. Acidic cleaners are usually those that contain ammonia or vinegar. Avoid getting glass cleaners on plumbing fixtures as some formulas can deteriorate the finish.

[Builder] Limited Warranty Guidelines

We will confirm that all mirrors are in acceptable condition during the orientation. [Builder] will correct scratches, chips, or other damage to mirrors noted during the orientation.

Paint and Stain

Homeowner Use and Maintenance Guidelines

Because of changes in the formula for paint (such as the elimination of lead to make paints safer), painted surfaces must be washed gently using mild soap and as little water as possible. Avoid abrasive cleaners, scouring pads, or scrub brushes. Flat paints show washing marks more easily than gloss paints do. Often better results come from touching up rather than washing the paint.

Colors

Your selection sheets are your record of the paint and stain color names, numbers, and brands in your home.

Exterior

Regular attention will preserve the beauty and value of your home. Check the painted and stained surfaces of your home's exterior annually. Repaint before much chipping or wearing away of the original finish occurs; this will save the cost of extensive surface preparation.

Plan on refinishing the exterior surface of your home approximately every two to three years or as often as your paint manufacturer suggests for your area and climate. Climatic conditions control the chemical structure of the paint used on the exterior. Over time, this finish will fade and dull a bit. Depending on the exposure to weather of each surface, the paint on some parts of your home may begin to show signs of deterioration sooner than others.

When you repaint the exterior of your home, begin by resetting popped nails and removing blistered or peeling portions of paint with a wire brush or putty knife. Sand, spot with primer, and then paint the entire area. Use a quality exterior paint formulated for local climate conditions.

Avoid having sprinklers spray water on the exterior walls of your home. This will cause blistering, peeling, splintering, and other damage to the home.

Severe Weather

Hail and wind can cause a great deal of damage in a severe storm, so inspect the house after such weather. Promptly report damage caused by severe weather to your insurance company.

Stain

For minor interior stain touch-ups, a furniture-polish-and-stain treatment is inexpensive, easy to use, and will blend in with the wood grain. Follow directions on the bottle.

Touch-Up

When doing paint touch-ups, use a small brush, applying paint only to the damaged spot. Touch-up may not match the surrounding area exactly, even if the same paint mix is used. When it is time to repaint a room, prepare the wall surfaces first by cleaning with a mild soap and water mixture or a reliable cleaning product.

We provide samples of each paint used on your home. Store these with the lids tightly in place and in a location where they are not subjected to extreme temperatures.

Wall Cracks

We suggest that you wait until after the first heating season to repair drywall cracks or other separations due to shrinkage.

See also Drywall.

[Builder] Limited Warranty Guidelines

During your orientation we will confirm that all painted or stained surfaces are in acceptable condition. [Builder] will touch up paint as indicated on the orientation list. You are responsible for all subsequent touch-up, except painting we perform as part of another warranty repair.

Cracking

As it ages, exterior wood trim will develop minor cracks and raised grain. Much of this will occur during the first year. Raised grain permits moisture to get under the paint and can result in peeling. This is not a defect in materials or workmanship. Paint maintenance of wood trim and gutters is your responsibility.

Fading

Expect fading of exterior paint or stain caused by the effects of sun and weather. [Builder] limited warranty excludes this occurrence.

Touch-Up Visible

Paint touch-up is visible under certain lighting conditions.

Wood Grain

Because of wood characteristics, color variations will result when stain is applied to wood. This is natural and requires no repair. Today's water-base paints often make wood grain visible on painted trim. [Builder] does not provide corrections for this condition.

Pests and Wildlife

Homeowner Use and Maintenance Guidelines

Insects such as ants, spiders, wasps, and bees, and animal life such as woodpeckers, squirrels, mice, and snakes, may fail to recognize that your home belongs to you. Addressing concerns involving these pests and wildlife goes with being a homeowner. Informational resources include, among others, the state wildlife service, animal control authorities, the county extension service, pest control professionals, Internet, and public library.

[Builder] Homeowner Manual

Phone Jacks

Homeowner Use and Maintenance Guidelines

Your home is equipped with telephone jacks as shown on the blueprints and selection sheets. Initiating phone service, additions to phone service, and moving phone outlets for decorating purposes or convenience are your responsibility.

[Builder] Limited Warranty Guidelines

[Builder] will correct outlets positioned so that a wall phone cannot be installed, for instance, if a kitchen phone outlet is positioned too close to a cabinet or countertop backsplash and prevents a wall phone from being connected.

[Builder] will repair wiring that does not perform as intended from the phone service box into the home. From the service box outward, care of the wiring is the responsibility of the local telephone service company.

Plumbing

Homeowner Use and Maintenance Guidelines

Your plumbing system has many parts, most of which require little maintenance. Proper cleaning, occasional minor attention, and preventive care will assure many years of good service from this system.

Aerators

Even though your plumbing lines have been flushed to remove dirt and foreign matter, small amounts of minerals may enter the line. Aerators on the faucets strain much of this from your water. Minerals caught in these aerators may cause the faucets to drip because washers wear more rapidly when they come in contact with foreign matter.

See also Dripping Faucet.

Basement Construction

If you perform any construction in your basement, ensure that the plumbing lines in the basement or crawl space are not isolated from the heating source without insulation being added.

Cleaning

Follow manufacturer's directions for cleaning fixtures. Avoid abrasive cleansers. They remove the shiny finish and leave behind a porous surface that is difficult to keep clean. Clean plumbing fixtures with a soft sponge and soapy water (a nonabrasive cleaner or a liquid detergent is usually recommended by manufacturers). Then polish the fixtures with a dry cloth to prevent water spots. Care for brass fixtures with a good-quality brass cleaner, available at most hardware stores.

Clogs

The main causes of toilet clogs are domestic items such as disposable diapers, excessive amounts of toilet paper, sanitary supplies, Q-tips, dental floss, and children's toys. Improper garbage disposal use also causes many plumbing clogs. Always use plenty of cold water when running the disposal. This recommendation also applies to grease; supplied with a steady flow of cold water, the grease congeals and is cut up by the blades. If you use hot water, the grease remains a liquid, then cools and solidifies in the sewer line. Allow the water to run 10 to 15 seconds after shutting off the disposal.

You can usually clear clogged traps with a plumber's helper (plunger). If you use chemical agents, follow directions carefully to avoid personal injury or damage to the fixtures.

Clean a plunger drain stopper—usually found in bathroom sinks—by loosening the nut under the sink at the back, pulling out the rod attached to the plunger, and lifting the stopper. Clean and return the mechanism to its original position.

Dripping Faucet

You can repair a dripping faucet by shutting off the water at the valve directly under the sink, then removing the faucet stem, changing the washer, and reinstalling the faucet stem. The shower head is repaired the same way. Replace the washer with another of the same type and size. You can minimize the frequency of this repair by remembering not to turn faucets off with excessive force. (Please note that some manufacturers do not use rubber washers.)

Extended Absence

If you plan to be away for an extended period, you should drain your water supply lines. To do this, shut off the main supply line and open the faucets to relieve pressure in the lines. You may also wish to shut off the water heater. Do this by turning off the cold water supply valve on top and the gas control at the bottom. Drain the tank by running a hose from the spigot on the bottom to the basement floor drain. If you leave the tank full, keep the pilot on and set the temperature to its lowest or "vacation" setting. Check manufacturer's directions for additional hints and instructions.

See also Extended Absence checklist.

Fiberglass Fixtures

For normal cleaning use a nonabrasive bathroom cleanser and sponge or nylon cleaning pad. Avoid steel wool, scrapers, and scouring pads. Auto wax can provide a shine and restore an attractive appearance.

Freezing Pipes

Provided the home is heated at a normal level, pipes should not freeze at temperatures above 0 degrees Fahrenheit. Set the heat at a minimum of 55 degrees F if you are away during winter months. Keep garage doors closed to protect plumbing lines running through this area from freezing temperatures.

In unusually frigid weather or if you will be gone more than a day or two, open cabinet doors to allow warm air to circulate around pipes. Use an ordinary hair dryer to thaw pipes that are frozen. Never use an open flame.

Gold or Brass Finish

Avoid using any abrasive cleaners on gold or antique brass fixtures. Use only mild detergent and water or a cleaning product recommended by the manufacturer.

Jetted Tubs

If your home includes a jetted tub follow manufacturer directions for its use and care. Never operate the jets unless the water level is at least one inch above the jets. Be cautious about using the tub if you are pregnant or have heart disease or high blood pressure; discuss the use of the tub with your doctor. Tie or pin long hair to keep it from away from the jets where it might become tangled—a potentially dangerous event.

Clean and disinfect the system every one to two months, depending on usage. To do this, fill the tub with lukewarm water and add one cup of liquid chlorine bleach. Run the jets for 10 to 15 minutes, drain and fill again. Run for 10 minutes with plain water, drain.

Auto wax will help seal and preserve your tub's surface. Avoid abrasive cleansers.

Laundry Tub

If you have a laundry room tub, the faucet does not have an aerator. This is to allow the laundry tub faucet to accept a hose connection.

Leaks

If a major plumbing leak occurs, the first step is to turn off the supply of water to the area involved. This may mean shutting off the water to the entire home. Then contact the appropriate contractor.

Low Flush Toilets

We want to draw your attention to a water-saving regulation that went into effect in 1993, which prohibits the manufacture of toilets that use more than 1.6 gallons of water per flush. In the search for a balance among comfort, convenience, and sensible use of natural resources, the government conducted several studies. The 1.6-gallon toilet turned out to be the size that overall consistently saves water.

As a result of implementing this standard, flushing twice is occasionally necessary to completely empty the toilet bowl. Even though you flush twice on occasion, rest assured that overall you are saving water and we have complied with the law. Similarly, flow restrictors are manufactured into most faucets and all shower heads and cannot be removed. We apologize for any inconvenience this may cause.

Low Pressure

Occasional cleaning of the aerators on your faucets (normally every three to four months) will allow proper flow of water. The water department controls the overall water pressure.

Main Shut-Off

The water supply to your home can be shut-off entirely in two locations. The first is at the street and the second is at the meter. We will point both of these out during your orientation.

Marble or Manufactured Marble

Marble and manufactured marble will not chip as readily as porcelain enamel but can be damaged by a sharp blow. Avoid abrasive cleansers or razor blades on manufactured marble; both damage the surface. Always mix hot and cold water at manufactured marble sinks; running only hot water can damage the sink.

Outside Faucets

Outside faucets (sillcocks) are freeze-proof, but in order for this feature to be effective, you must remove hoses during cold weather, even if the faucet is located in your garage. If a hose is left attached, the water that remains in the hose can freeze and expand back into the pipe, causing a break in the line. Repair of a broken line that feeds an exterior faucet is a maintenance item. Note that [Builder] does not warrant sillcocks against freezing.

Porcelain

You can damage porcelain enamel with a sharp blow from a heavy object or by scratching. Do not stand in the bathtub wearing shoes unless you have placed a protective layer of newspaper over the bottom of the tub. If you splatter paint onto the porcelain enamel surfaces during redecorating, wipe it up immediately. If a spot dries before you notice it, use a recommended solvent.

Running Toilet

To stop running water, check the shut-off float in the tank. You will most likely find it has lifted too high in the tank, preventing the valve from shutting off completely. In this case, gently bend the float rod down until it stops the water at the correct level. The float should be free and not rub the side of the tank or any other parts. Also check the chain on the flush handle. If it is too tight, it will prevent the rubber stopper at the bottom of the tank from sealing, resulting in running water.

Shut-Offs

Your main water shut-off is located near your meter. You use this shut-off for major water emergencies such as a water line break or when you install a sprinkler system or build an addition to your home. Each toilet has a shut-off on the water line under the tank. Hot and cold shut-offs for each sink are on the water lines under the sink.

Sprinklers

You should routinely inspect sprinkler heads and provide seasonal service to maintain proper functioning.

See also Landscaping/Sprinkler.

Stainless Steel

Clean stainless steel sinks with soap and water to preserve their luster. Avoid using abrasive cleaners or steel wool pads; these will damage the finish. Prevent bleach from coming into prolonged contact with the sink as it can pit the surface. An occasional cleaning with a good stainless steel cleaner will enhance the finish. Rub in the direction of the polish or grain lines and dry the sink to prevent water spots.

Avoid leaving produce on a stainless steel surface, since prolonged contact with produce can stain the finish. Also avoid using the sink as a cutting board; sharp knives will gouge the finish.

Local water conditions affect the appearance of stainless steel. A white film can develop on the sink if you have over-softened water or water with a high concentration of minerals. In hard water areas, a brown surface stain can form appearing like rust.

Tank Care

Avoid exposing the toilet to blows from sharp or heavy objects, which can cause chipping or cracking. Avoid abnormal pressures against the sides of the tank. It is possible to crack the tank at the points where it is attached to the bowl.

Water Filter or Softener

If you install either a water filter or a water softener, carefully read the manufacturer's literature and warranty for your specific model.

If your home includes a septic system, prior to installing a water softener, discuss with the vendor whether the system you are considering will adversely affect your septic system.

See also Septic System.

TROUBLESHOOTING TIPS: PLUMBING

No Water Anywhere in the Home

Before calling for service, check to confirm that the:

- Main shut off on the meter inside your home is open.
- Main shut off at the street is open.
- Individual shut-offs for each water-using item are open.

No Hot Water

See Water Heater

Leak Involving One Sink, Tub, or Toilet

- Check caulking and grout.
- Confirm shower door or tub enclosure was properly closed.
- Turn water supply off to that item.
- Use other facilities in your home and report problem on next business day.

Leak Involving a Main Line

- Turn water off at the meter in your home.
- Call emergency number for service.

Back Up at One Toilet

If only one toilet is affected, corrections occur during normal business hours.

- Shut off the water supply to the toilet involved.
- Use a plunger to clear the blockage.
- Use a snake to clear the blockage.
- If you've been in your home fewer than 30 days, contact [Builder] or the plumber listed on your Emergency Phone Numbers sheet.
- If you've been in your home over 30 days, contact a router service.

Sewer Back Up Affecting Entire Home

- If you've been in your home fewer than 30 days, contact [Builder] or the plumber listed on your Emergency Phone Numbers sheet.
- If you've been in your home over 30 days, contact a router service.
- Remove personal belongings to a safe location. If items are soiled, contact your homeowner insurance company.

Even if the troubleshooting tips do not identify a solution, the information you gather will be useful to the service provider you call.

[Builder] Limited Warranty Guidelines

During the orientation we will confirm that all plumbing fixtures are in acceptable condition and are functioning properly, and that all faucets and drains operate freely.

Clogged Drain

[Builder] will correct clogged drains that occur during the first 30 days after closing. If a household item is removed from a clogged drain during this time, we will bill you for the drain service. After the first 30 days, you are responsible for correcting clogged drains.

Cosmetic Damage

[Builder] will correct any fixture damage noted on the orientation list. Repairing chips, scratches, or other surface damage noted subsequent to the orientation list is your responsibility.

Exterior Faucets

[Builder] will repair leaks at exterior faucets noted on the orientation list. Subsequent to orientation, repair of a broken line to an exterior faucet is your responsibility.

Freezing Pipes

Provided the home is heated at a normal level, pipes should not freeze. Set heat at 55 degrees F if you are away during winter months. Keep garage doors closed to protect plumbing lines that run through this area.

Leaks

[Builder] will repair leaks in the plumbing system. If a plumbing leak caused by a warranted item results in drywall or floor covering damage, [Builder] will repair or replace items that were part of the home as originally purchased. We do not make adjustments for secondary damages (for example, damage to wallpaper, drapes, and personal belongings). Insurance should cover these items.

Noise

Changes in temperature or the flow of the water itself will cause some noise in the pipes. This is normal and requires no repair. [Builder] will repair persistent water hammer. Expect temperatures to vary if water is used in more than one location in the home at the exact same time.

Supply

[Builder] will correct construction conditions that disrupt the supply of water to your home if they involve service from the main water supply to your home, provided actions of yours have not caused the problem. Disruption of service due to failure of the water department system is the responsibility of the water department to correct.

Property Boundaries

Homeowner Use and Maintenance Guidelines

At closing you will receive a copy of a survey that shows your lot and the location of your home on the lot. To construct the home [Builder] established the property boundaries and corners.

During construction, some of the monuments that mark the lot corners may be affected or covered up by grading, excavation, installation of utility lines and other typical construction activities. If you wish to install a fence, swimming pool, add a deck or patio to your home, or otherwise establish a permanent structure, we advise that you have professional surveyors locate and mark property boundaries to be certain they are accurate and you have found all corners.

See also Easements.

Railings

Homeowner Use and Maintenance Guidelines

Stained or wrought iron railings in your home require little maintenance beyond occasional dusting or polishing. Protect railings from sharp objects or moisture. Cover them during move-in so large pieces of furniture do not cause dents or scratches.

Stained railings will show variation in the way the wood grain took the stain. Some designs show seams where pieces of wood came together to form the railing.

[Builder] Limited Warranty Guidelines

During the orientation we will confirm that all railings are in good condition. [Builder] installs railings in positions and locations to comply with applicable building codes. Railings should remain securely attached with normal use.

Resilient Flooring

Homeowner Use and Maintenance Guidelines

Although resilient floors are designed for minimum care, they do have maintenance needs. Follow any manufacturer's specific recommendations for care and cleaning. Some resilient floors require regular application of a good floor finish. This assures you of retaining a high gloss. However, avoid using cleaning or finishing agents on the new floor until the adhesive has thoroughly set. This will take about two weeks.

Color and Pattern

Your color selection sheets provide a record of the brand, style, and color of floor coverings in your home. Please retain this information for future reference.

Limit Water

Wipe up spills and vacuum crumbs instead of washing resilient floors frequently with water. Limit mopping or washing with water; excessive amounts of water on resilient floors can penetrate seams and get under edges, causing the material to lift and curl.

Moving Furniture

Moving appliances across resilient floor covering can result in tears and wrinkles. Install coasters on furniture legs to prevent permanent damage. If you damage the resilient floor, you can have it successfully patched by professionals. If any scraps remain when installation of your floor covering is complete, we leave them in the hope that having the matching dye lot will make such repairs less apparent.

No-Wax Flooring

The resilient flooring installed in your home is the no-wax type. No-wax means a clear, tough coating that provides both a shiny appearance and a durable surface. However, even this surface will scuff or mark. Follow the manufacturer's recommendations for maintaining the finish.

Raised Nail Heads

Raised nail heads are the result of movements of the floor joist caused by natural shrinkage and deflection. We have used special nails and glued the underlayment to help minimize this movement. If a nail head becomes visible through resilient flooring, place a block of wood over it and hit the block with a hammer to reset the nail.

Scrubbing and Buffing

Frequent scrubbing or electric buffing is harder on floors than regular foot traffic. Use acrylic finishes if you scrub or buff.

Seams

Any brand or type of resilient flooring may separate slightly due to shrinkage. Seams can lift or curl if excessive moisture is allowed to penetrate them. You can use a special caulking at tub or floor joints to seal seams at those locations. Avoid getting large amounts of water on the floor from baths and showers.

[Builder] Limited Warranty Guidelines

We will confirm that resilient floor covering is in acceptable condition during your orientation. [Builder] limited warranty does not cover damage to resilient floors caused by moving furniture or appliances into the home. We can assist you in contacting professionals who can repair such damage if it occurs in your home. [Builder] is not responsible for discontinued selections.

Adhesion

Resilient floor covering should adhere. [Builder] will repair lifting or bubbling and nail pops that appear on the surface.

[Builder] Homeowner Manual

Ridges

[Builder] has sanded and filled the joints of underlayment to minimize the possibility of ridges showing through resilient floor coverings. Ridging is measured by centering a 6-inch straight edge perpendicular to the ridge with one end tight to the floor. If the opposite end of the straight edge is 1/8 inch or more from the floor, [Builder] will repair this condition.

Seams

Seams will occur and are sealed at the time of installation. [Builder] will correct gaps in excess of 1/16 inch where resilient flooring pieces meet or 1/8 inch where resilient flooring meets another material. [Builder] will correct curling at seams unless caused by excessive water.

Roof

Homeowner Use and Maintenance Guidelines

The shingles on your roof do not require any treatment or sealer. The less activity your roof experiences, the less likely it is that problems will occur.

Clean Gutters

Maintain the gutters and downspouts so that they are free of debris and able to quickly drain precipitation from the roof.

Ice Dam

On occasion, depending on conditions and exposure, as rising heat from inside your home melts snow on the roof, the water runs down and when it reaches the cold eaves, it may freeze. An accumulation of this type of ice dams the subsequent runoff and the water begins to back up, sometimes working its way up and under shingles, ultimately leading into you home through windows or ceilings.

If your home design or orientation makes it vulnerable to this occurrence, you may want to install an electric gutter heater strip in the susceptible areas.

Leaks

If a leak occurs, try to detect the exact location. This will greatly simplify finding the area that requires repair when the roof is dry.

Limit Walking

Limit walking on your roof. Your weight and movement can loosen the roofing material and in turn result in leaks. Never walk on the roof of your home when the shingles are wet—they are slippery.

Severe Weather

After severe storms, do a visual inspection of the roof for damages. Notify your insurance company if you find pieces of shingle in the yard or shingle edges lifted on the roof.

TROUBLESHOOTING TIPS: ROOF LEAK

Please keep in mind that roof leaks cannot be repaired while the roof is wet. However, you can get on the schedule to be in line when conditions dry out, so do call in your roof leak.

- Confirm the source of the water is the roof rather than from a
    - —Plumbing leak
    - —Open window on a higher floor
    - —Ice dam
    - —Clogged gutter or downspout
    - —Blowing rain or snow coming in through code required roof vents
    - —Gap in caulking
- Where practical, place a container under dripping water.
- If a ceiling is involved, use a screwdriver to poke a small hole in the drywall to release the water.
- Even if the troubleshooting tips do not identify a solution, the information you gather will be useful to the service provider you call.
- Remove personal belongings to prevent damage to them. If damage occurs, contact your homeowner insurance company to submit a claim.
- Report the leak to [Builder] during first available business hours.

[Builder] Limited Warranty Guidelines

[Builder] will repair roof leaks other than those caused by severe weather, such as hail damage, or some action you have taken, such as walking on the roof. Roof repairs are made only when the roof is dry.

Ice Dam

An ice build-up (ice dam) may develop in the eaves during extended periods of cold and snow. Your homeowner insurance may cover this damage which is excluded from warranty.

Inclement Weather

Storm damage is excluded from warranty coverage. Notify your homeowner insurance company if storm damage is discovered.

Rough Carpentry

[Builder] Limited Warranty Guidelines

Some floor and stair squeaks are unavoidable. Although [Builder] does not warrant against floor squeaks, a reasonable effort will be made to correct them.

Floor Deflection

Floors will deflect (bend) when walked on. This will be more noticeable next to hutches, bookcases, pianos, chairs, and other heavy furniture. This is not a structural deficiency and [Builder] will take no action for this occurrence.

Floor Level

Floors will be level to within 1/4 inch within any 32-inch distance as measured perpendicular to any ridge or indentation. [Builder] will correct floor slope that exceeds 1/240 of the room.

Plumb Walls

[Builder] will correct walls that are out of plumb more than 1/2 inch in an 8-foot distance or walls that are bowed more than 1/4 inch in any 32-inch measurement.

Septic System

Homeowner Use and Maintenance Guidelines

A septic system consist of two basic parts. First a septic tank, and second an underground disposal field. Bacteria break down solids forming a sludge which is moved by incoming water out to the disposal field where is filtered out into the soil. To help preserve the effectiveness of the system, keep these points in mind:

- Avoid disposing of chemicals such as solvents, oils, points, and so on, through the septic system
- Avoid using commercial drain cleaners. They can kill the bacteria that are working to break down the solid waste matter.
- Food from a disposal decomposes more slowly and adds to the solids in the tank. Coffee grounds may clog the system.

- Avoid disposing of any paper product (diapers, sanitary supplies, paper towels and so on) other than toilet paper through the system.
- Do not rely on yeast or chemical additive to digest sludge. They are not an alternative to regular pumping and may actually harm the system.
- Drain surface water away from the disposal field. Eliminate unnecessary sources of water in the area of the disposal field. Plant only sod over the disposal field. Avoid fertilizers in this area.
- Conserve indoor water use to put less strain on the system. Correct leaky faucets or running toilets promptly. Keep in mind that a water softener will generate 30 to 85 gallons of water every regeneration cycle.
- Do not drive on the disposal field or build over it.

Pumping the System

Over time, the matter not broken down by the bacteria can clog the system. This will happen in spite of careful use and good maintenance. To prevent serious problems, regular pumping to clean out the tank is essential—usually every 1 to 2 years, more often if usage is heavy.

System Failure

Signs that your septic system is failing include:

- Black water with a foul odor backing up in drains or toilets.
- Toilets flush slowly.
- Water ponds on top of the disposal field.
- Grass stays green over the disposal field even in dry weather.

If you believe your system requires attention, call a professional to assess the situation. Have the system pumped. If a new system is required, a permit must be obtained from the county or municipality where your home is located.

Water Softener

Prior to installing a water softener, discuss with the vendor whether the system you are considering will adversely affect your septic system.

[Builder] Limited Warranty Guidelines

During the orientation we confirm that the septic system is working properly and that you are familiar with the location of the tank and disposal field.

While we install the system in accordance with codes and plans based on your soil conditions, we do not warrant that the septic system will function indefinitely. Weather, ground water, environmental conditions, topography, as well as your family's habits can all generate unpredictable effects.

Shower Doors or Tub Enclosures

Homeowner Use and Maintenance Guidelines

Shower doors and tub enclosures require minimal care. Using a squeegee to remove water after a bath or shower will keep mineral residue and soap film to a minimum. A coating of wax can also help prevent build up of minerals and soap.

Use cleaning products suggested by the manufacturer to avoid any damage to the trim and hardware.

Avoid hanging wet towels on corners of doors; the weight can pull the door out of alignment and cause it to leak.

Check and touch-up caulking on an as needed basis.

[Builder] Limited Warranty Guidelines

During your orientation we will confirm the good condition of all shower doors and tub enclosures. [Builder] warrants that shower doors and tub enclosures will function according to manufacturer specifications.

Siding

Homeowner Use and Maintenance Guidelines

Siding expands and contracts in response to changes in humidity and temperature. Slight waves are visible in siding under moist weather conditions; shrinkage and separations will be more noticeable under dry conditions. These behaviors cannot be entirely eliminated.

Wood and Wood Products

Wood or wood-product siding will require routine refinishing. The timing will vary with climatic conditions. Maintain caulking to minimize moisture entry into the siding. Note that some paint colors will require more maintenance than others and some sides of the home may show signs of wear sooner based on their exposure to the elements. Some wood siding, such as cedar, is subject to more cracking and will require more maintenance attention.

Vinyl

Vinyl siding will occasionally require cleaning. Start at the top to avoid streaking and use a cleaning product recommended by your siding manufacturer. Follow directions carefully.

Cement Based Products

Cement based siding will require repainting and caulking just as wood products do.

See also Paint and Wood Trim.

[Builder] Limited Warranty Guidelines

[Builder] warrants all siding to be free of defects in material and workmanship. We will confirm the good condition of the siding during your orientation. Subsequent damage to the siding will by your responsibility to repair.

[Builder] will caulk and apply touch-up paint to cracks that exceed 3/16 inch. We provide this repair one time only near the end of the first year. Paint or stain touch-up will not match.

We will correct any separation at joints or where siding meets another material if the separation allows water to enter the home. [Builder] will correct delaminating siding.

Smoke Detectors

Homeowner Use and Maintenance Guidelines

Read the manufacturer's manual for detailed information on the care of your smoke detectors.

Battery

If a smoke detector makes a chirping sound that is a sign that the battery needs to be replaced. Follow manufacturer instructions for installing a new battery. Most smoke detectors use a 9 volt battery.

Cleaning

For your safety, clean each smoke detector monthly to prevent a false alarm or lack of response in a fire. After cleaning, push the test button to confirm the alarm is working.

Locations

Smoke detectors are installed in accordance with building codes, which dictate locations. [Builder] cannot omit any smoke detector and you should not remove or disable any smoke detector.

[Builder] Homeowner Manual

[Builder] Limited Warranty Guidelines

[Builder] does not represent that the smoke detectors will provide the protection for which they are installed or intended. We will test smoke detectors during the orientation to confirm that they are working and to familiarize you with the alarm. You are responsible for obtaining fire insurance.

Stairs

Homeowner Use and Maintenance Guidelines

No known method of installation prevents all vibration or squeaks in a staircase. A shrinkage crack will develop where the stairs meet the wall. When this occurs, apply a thin bead of latex caulk and, when dry, touch up with paint.

[Builder] Limited Warranty Guidelines

Although [Builder] does not warrant against stair vibration and squeaks, a reasonable effort will be made to correct them.

Stucco

Homeowner Use and Maintenance Guidelines

Stucco is a brittle cement product that is subject to expansion and contraction. Minor hairline cracks will develop in the outer layer of stucco. This is normal and does not reduce the function of the stucco in any way.

Drainage

To ensure proper drainage, keep dirt and concrete flatwork a minimum of 6 inches below the stucco screed (mesh underneath final coat of stucco). Do not pour concrete or masonry over the stucco screed or right up to the foundation.

Efflorescence

The white, powdery substance that sometimes accumulates on stucco surfaces is called efflorescence. This is a natural phenomenon and cannot be prevented. In some cases, you can remove it by scrubbing with a stiff brush and vinegar. Consult your home center or hardware store for commercial products to remove efflorescence.

Sprinklers

Since stucco is not a water barrier, avoid spraying water from irrigation or watering systems on stucco surfaces to avoid possible leaks. Check the spray from the lawn and plant irrigation system frequently to make certain that water is not spraying or accumulating on stucco surfaces.

[Builder] Limited Warranty Guidelines

One time during the warranty period, [Builder] will repair stucco cracks. The repair will not exactly match the surrounding area.

Sump Pump

Homeowner Use and Maintenance Guidelines

If conditions on you lot made it appropriate, the foundation design includes a perimeter drain and sump pump. The perimeter drain runs around the foundation to gather water and channel it to the sump pit, or crock. When the water reaches a certain level, the pump comes on and pumps the water out of your home. Read and follow the manufacturer's directions for use and care of your sump pump.

Continuous Operation

The pump may run often or even continuously during a heavy storm or long periods of rain. This is normal under such conditions.

Discharge

Know where the discharge for your sump pump system is and keep the end of the drain clear of debris so that water can flow out easily.

Power Supply

The sump pump runs on electricity. If power goes off, the pump cannot operate. Storm water (not sewage) could then enter your basement. You may wish to install a back-up system to guard against this possibility. Homeowner insurance does not usually cover damage to your property from this source; you may want to obtain a rider to cover this.

Roof Water

Ensure that roof water drains quickly away from the home to avoid circulating it through your sump pump. Keep downspout extensions or splash blocks in place to channel water away from your home.

[Builder] Homeowner Manual

Routine Check

Periodically check to confirm the pump is plugged in, the circuit breaker is on and that the pump operates. To check the operation of your sump pump, pour five gallons of water into the sump pump crock (hole). The pump should come on and pump the water out. Follow this procedure once a year.

Trees and Shrubs

Avoid planting trees or shrubs with aggressive root growth patterns near your home's foundation. The roots can make their way into the perimeter drain and eventually clog the system.

[Builder] Limited Warranty Guidelines

During your orientation we will discuss the sump pump and confirm it is operational. The pump is classified as an appliance and is warranted by the manufacturer.

Swimming Pools

Homeowner Use and Maintenance Guidelines

If your home includes a swimming pool, be aware of important safety and care requirements. Local ordinances require that you secure the pool area with a fence and locked gate to prevent unauthorized entry and use of your pool. Establish safe practices with children regarding proper pool behaviors and circumstances under which they can enter the water.

Chemicals

Carefully study and follow information regarding the pool's chemistry. You are responsible for supplying all appropriate chemical treatments.

Cleaning

Regular cleaning of the pool's surfaces is essential for comfortable and healthy enjoyment. Keep glass and debris out of the pool area.

Filters and Pumps

Maintain the pool filters and pumps according to each manufacturer's directions.

Professional Services

Consider retaining the services of a professional pool service to clean the pool, maintain the systems, and treat the water.

[Builder] Limited Warranty Guidelines

During the orientation, we will confirm that all pool surfaces are in acceptable condition. Repair of any surface damage noted subsequent to that is your responsibility. The pool installer will set a separate appointment with you to instruct you in the use and care of equipment and review chemical treatment of the pool water.

Pool equipment should function as designed provided you follow all maintenance steps.

Termites

Homeowner Use and Maintenance Guidelines

We treat the foundation of your home for termites and provide you with a certificate confirming that treatment. Plan to renew this treatment annually or as directed by the literature that accompanies the certificate. Treatment for other types of insects or animal infestations is your responsibility.

Regular Inspections

Regularly inspect your home for signs of termites or conditions that would allow their attack.
- Check for wrinkles or waves in wood trim.
- Tap wood to see if it sound or feels hollow.
- Inspect under the carpet tack strip by lifting the edge of carpet in the corner of a room. The tack strip is untreated and provides a convenient path for termites through your home.
- Watch for tubes of dirt, called mud tubes, that extend from the soil up to your home.
- Keep soil away from any wood parts of your home.
- Be certain all roof water and precipitation moves quickly away from your home's foundation.
- Avoid storing wood on the ground and against your home.
- Maintain a safe zone of at least two feet in width around the perimeter of your home. Avoid planting grass or shrubs, installing any sprinkler device, of digging of any kind in this area. If you disturb this area, have it re-treated to restore protection.
- Before installing stepping stones, river rock, concrete, or so on, against the home, chemically treat the area that will be underneath the new material.
- If you add onto or change the exterior of your home, be sure to have the areas treated first.

If you believe you see signs of termites or if you have any questions, contact your termite treatment company for guidance.

[Builder] Limited Warranty Guidelines

We certify treatment of your foundation for termites at closing. This is our final action for termites. [Builder] warranty excludes treatment for any other insect (such as ants) or animal (such as mice) infestations.

[Builder] Homeowner Manual

Ventilation

Homeowner Use and Maintenance Guidelines

Homes today are built more tightly than ever. This saves energy dollars but creates a potential concern. Condensation, cooking odors, indoor pollutants, radon, and carbon monoxide may all accumulate. We provide mechanical and passive methods for ventilating homes. Your attention to ventilation is important to health and safety. Building codes require attic and crawl space vents to minimize accumulation of moisture.

Attic Vents

Attic ventilation occurs through vents in the soffit (the underside of the overhangs) or on gable ends. Driving rain or snow sometimes enters the attic through these vents. Do not cover them to prevent this. Instead, cover the insulation in front of the vent. When you do this, precipitation that blows in safely evaporates and ventilation can still occur.

Crawl Space Vents

Homes with crawl spaces usually include two or more vents. Open crawl space vents for summer months and close them for winter months, pulling insulation over them. Failure to close these vents and replace insulation may result in plumbing lines freezing in the crawl space. This occurrence is not covered by your warranty.

Daily Habits

Your daily habits can help keep your home well ventilated:

- Do not cover or interfere in any way with the fresh air supply to your furnace.
- Develop the habit of running the hood fan when you are cooking.
- Ditto the bath fans when bathrooms are in use.
- Air your house by opening windows for a time when weather permits.

Proper ventilation will prevent excessive moisture from forming on the inside of the windows. This helps reduce cleaning chores considerably.

[Builder] Limited Warranty Guidelines

[Builder] warranty guidelines for active components (for example, exhaust fans) are discussed under the appropriate headings (such as electrical systems, heating system, and so on).

[Builder] Homeowner Manual

Water Heater: Electric

Homeowner Care and Maintenance

Carefully read the manufacturer's literature and warranty for your specific model of water heater.

Drain Tank

Review and follow the manufacturer's timetable and instructions for draining several gallons of water from the bottom of the water heater. This reduces build-up of chemical deposits from the water, thereby prolonging the life of the tank as well as saving energy dollars. Also drain the tank if it is being shut down during periods of freezing temperatures. Carefully follow the instructions in the manufacturer's literature.

Element Cleaning or Replacement

The heating elements in the water heater will require periodic cleaning. The frequency is determined in part by the quality of the water in your area. Again, refer to the manufacturer's literature for step-by-step instructions and drawings, or contact an authorized service company.

Pressure Relief Valve

At least once each year, manually operate the pressure relief valve. Stay clear of the discharge line to avoid injury. See manufacturer's literature for diagrams and detailed instructions.

Safety

Keep the area around a water heater clear of stored household items. Never use the top of the water heater as a storage shelf.

Temperature

Temperature settings on an electric water heater will produce approximately the temperatures listed below:

Hot	120 degrees F
A	130 degrees F
B	140 degrees F
C	150 degrees F
Very Hot	160 degrees F

The recommended setting for operation of a dishwasher is B, or 140 degrees. Higher settings can waste energy dollars and increase the danger of injury from scalding. Hot water will take longer to arrive at sinks, tubs, and showers that are farther from the water heater.

[Builder] Homeowner Manual

TROUBLE SHOOTING TIPS: NO HOT WATER

Before calling for service, check to confirm that the

- Water heater breaker on your main electric panel is in the on position. (Remember if a breaker trips you must turn it from the tripped position to the off position before you can turn it back on.)
- Temperature setting is not on "vacation" or too low.
- Water supply valve is open.

Refer to the manufacturer's literature for specific locations of these items and possibly other troubleshooting tips.

Even if the trouble shooting tips do not identify a solution, the information you gather will be useful to the service provider you call.

[Builder] Limited Warranty

Refer to the manufacturer's limited warranty for complete information regarding warranty coverage on your water heater.

Water Heater: Gas

Homeowner Use and Maintenance Guidelines

Carefully read and follow the manufacturer's literature for your specific model of water heater.

Condensation

Condensation inside your new water heater may drip onto the burner flame. This causes no harm and in most cases will disappear in a short period of time.

Drain Tank

Review and follow manufacturer's timetable and instructions for draining several gallons of water from the bottom of the water heater. This reduces the build-up of chemical deposits from the water, prolonging the life of the tank and saving energy dollars.

Pilot

Never light a gas pilot when the water heater tank is empty. Always turn off the gas before shutting off the cold water supply to the tank.

To light the water heater pilot, first remove the cover panel on the tank to expose the pilot. Then rotate the on-off-pilot knob to the pilot position. When the knob is in this position, the red button can be depressed.

While depressing the red button, hold a match at the pilot. Once the pilot lights, continue to hold the red button down for 30 to 60 seconds. When you release the red button, the pilot should stay lit. If it does not, wait several minutes to allow the gas to dissipate from the tank and repeat the entire process. If it stays lit, rotate the on-off pilot knob to the on position.

Reinstall the cover panel and then adjust the temperature setting with the regulating knob on the front of the tank.

Water heaters sometimes collect small quantities of dirty water and scale in the main gas lines, which may put out the pilot light.

While away from home for an extended period of time, set the temperature to its lowest point and leave the pilot lit.

Safety

Vacuum the area around a gas-fired water heater to prevent dust from interfering with proper flame combustion. Avoid using the top of a heater as a storage shelf.

Temperature

The recommended thermostat setting for normal everyday use is "normal." Higher settings can result in wasted energy dollars and increase the danger of injury from scalding. Hot water will take longer to arrive at sinks, tubs, and showers that are farther from the water heater.

TROUBLESHOOTING TIPS: NO HOT WATER

Before calling for service, check to confirm that the:

- Pilot is lit. (Directions will be found on the side of the tank.)
- Temperature setting is not on "vacation" or too low.
- Water supply valve is open.

Refer to the manufacturer's literature for specific locations of these items and possibly other troubleshooting tips.

Even if the trouble shooting tips do not identify a solution, the information you gather will be useful to the service provider you call.

[Builder] Limited Warranty Guidelines

Refer to the manufacturer's limited warranty for information regarding coverage of the water heater.

See also Plumbing

Windows, Screens, and Sliding Glass Doors

Homeowner Use and Maintenance Guidelines

Contact a glass company for reglazing of any windows that break. Glass is difficult to install without special tools.

Acrylic Block

Clean during moderate temperatures with only a mild soap and warm water using a sponge or soft cloth and dry with a towel. Avoid abrasive cleaners, commercial glass cleaner, razors, brushes, or scrubbing devices of any kind. Minor scratches can often be minimized using by rubbing a mild automotive polish.

Aluminum

Clean aluminum metal surfaces with warm, clear water. Do not use powdered cleaner. After each cleaning, apply a silicone lubricant. Clean glass as needed with vinegar and water, a commercial glass cleaner, or the product recommended by the window manufacturer.

Condensation

Condensation on interior surfaces of the window and frame is the result of high humidity within the home and low outside temperatures. Your family's lifestyle controls the humidity level within your home. If your home includes a humidifier, closely observe the manufacturer's directions for its use.

Screen Storage and Maintenance

Many homeowners remove and store screens for the winter to allow more light into the home. To make re-installation more convenient, label each screen as you remove it. Use caution: screens perforate easily and the frames bend if they are not handled with care. Prior to re-installing the screen, clean them with a hose and gentle spray of water.

Sills

Window sills in your home are made of wood, wood product, man-made marble, or marble. The most common maintenance activity is dusting. Twice a year, check caulking and touch-up as needed. Wax is not necessary but can be used to make sills gleam. Protect wood and wood product sills from moisture. If you arrange plants on a sill, include a plastic tray under the pot.

Sliding Glass Doors

Sliding glass doors are made with tempered glass which is more difficult to break than ordinary glass. If broken, tempered glass breaks into small circular pieces rather than large splinters which can easily cause injury.

Keep sliding door tracks clean for smooth operation and to prevent damage to the door frame. Silicone lubricants work well for these tracks. Acquaint yourself with the operation of sliding door hardware for maximum security.

Under certain lighting conditions, door glass may be hard to see. If you keep the screen fully closed when the glass door is open, your family will be accustomed to opening something before going through. You may want to apply a decal to the glass door to make it readily visible.

Sticking Windows

Most sliding windows (both vertical and horizontal) are designed for a 10-pound pull. If sticking occurs or excessive pressure is required to open or close a window, apply a silicone lubricant. This is available at hardware stores. Avoid petroleum-based products.

Tinting

Applying tinting of foil lining to dual pane windows can result in broken windows due to heat build-up. Some manufacturers void their warranty on the windows if you apply tinting or foil lining. Contact the manufacturer to check on their current policy before you apply such coatings.

Weep Holes

In heavy rains, water may collect in the bottom channel of window frames. Weep holes are provided to allow excess water to escape to the outside. Keep the bottom window channels and weep holes free of dirt and debris for proper operation.

[Builder] Limited Warranty Guidelines

We will confirm that all windows, screens, and sliding glass doors are in acceptable condition during the orientation. [Builder] will repair or replace broken windows or damaged screens noted on the orientation list. Windows should operate with reasonable ease and locks should perform as designed. If they do not, [Builder] will provide adjustments.

Condensation

Condensation on interior surfaces of the window and frame is the result of high humidity within the home and low outside temperatures. You influence the humidity level within your home; [Builder] provides no corrective measure for this condition.

Condensation that accumulates between the panes of glass in dual-glazed windows indicates a broken seal. [Builder] will replace the window if this occurs during the warranty period.

Infiltration

Some air and dust will infiltrate around windows, especially before the installation of landscaping in the general area. [Builder] warranty excludes this occurrence.

Scratches

[Builder] confirms that all window glass is in acceptable condition at the orientation. Minor scratches on windows can result from delivery, handling, and other construction activities. [Builder] will replace windows that have scratches readily visible from a distance of 4 feet. [Builder] does not replace windows that have scratches visible only under certain lighting conditions.

Tinting

If you add tinting to dual-glazed windows, all warranties are voided. Damage can result from condensation or excessive heat build-up between the panes of glass. Refer to the manufacturer's literature for additional information.

See also Ventilation

Wood Trim

Homeowner Use and Maintenance Guidelines

Shrinkage of wood trim occurs during the first two years or longer, depending on temperature and humidity. All lumber is more vulnerable to shrinkage during the heating season. Maintaining a moderate and stable temperature helps to minimize the effects of shrinkage. Wood will shrink less lengthwise than across the grain. Wood shrinkage can result in separation at joints of trim pieces. You can usually correct this with caulking and touch-up painting.

Shrinkage may also cause a piece of trim to pull away from the wall. If this occurs, drive in another nail close to, but not exactly in, the existing nail hole. Fill the old nail hole with putty and touch up with paint as needed. If the base shoe (small trim between base molding and the floor) appears to be lifting from the floor, this is probably due to slight shrinkage of the floor joists below. Again, you can correct this condition by removing the old nails and renailing. You may prefer to wait until after the first heating season to make any needed repairs at one time when redecorating.

See also Expansion and Contraction

[Builder] Limited Warranty Guidelines

During the orientation we will confirm that wood trim is in acceptable condition. Minor imperfections in wood materials will be visible and will require no action. [Builder] will correct readily noticeable construction damage such as chips and gouges listed during the orientation.

Exterior

[Builder] will caulk and apply touch-up paint to cracks in exterior trim components that exceed 3/16 inch. We provide this repair one time only near the end of the first year. Paint or stain touch-up will not match. We will correct any separation at joints that allows water to enter the home.

Raised Grain

Because of the effects of weather on natural wood, you should expect raised grain to develop. This is normal and not a defect in the wood or paint. Warranty coverage excludes this condition.

[Builder Logo]

Warranty Service Request

For your protection and to allow efficient operations, our warranty service system is based on your written report of nonemergency items. Please use this form to notify us of warranty items. Mail to the address shown above. We will contact you to set an inspection appointment. Service appointments are available from 7:00 a.m. to 4:00 p.m., Monday through Friday. *Thank you for your cooperation.*

Name _____ Date_____
Address _____ Community_____
Ph (Home) _____ Lot #_____
Ph (Work) _____ Plan_____
Ph (Work) _____ Closing Date_____

		Service Action*		
Location	**Service Requested**	Warranty	Courtesy	Maintenance

*Warranty or Courtesy indicates a [Builder] responsibility. Maintenance indicates a homeowner responsibility.

Comment:_____
_____
_____

Homeowner _____

[Builder Logo]

One-Time Repairs

We provide several first-time repairs for your home. Your Homeowner Manual lists these under individual headings such as drywall and grout in the Caring for Your Home section. We provide this service as a courtesy and to give you an opportunity to observe methods and materials needed for ongoing maintenance of your home.

Only *one* one-time repair request per home during the warranty period, please. We suggest sending this in near the end of your warranty year to maximize the benefits you receive. Simply complete and mail or fax this form to our office with your year-end warranty list. Thank you!

Room	*Location*

Comment: _____

_____

_____

_____

Homeowner _____

[Builder Logo]

Dear Homeowner,

We want our Homeowner Manual to be relevant and useful to the needs of our customers and homeowners. We revise this material once each year and would appreciate your feedback and comments.

1. Please indicate how you used this manual:

   ☐ Read it from cover to cover
   ☐ Briefly looked it over
   ☐ Looked up answers to specific questions on occasion
   ☐ Did not use it at all

1. Did you find the information:
   ☐ Useful
   ☐ Easy to understand
   ☐ Accurate

2. What sections were most helpful?
   ☐ Purchasing Your Home
   ☐ Arranging Your Loan
   ☐ New Home Selections
   ☐ Construction of Your Home
   ☐ Homeowner Orientation
   ☐ Closing on Your Home
   ☐ Caring for Your Home

3. What topics should we add?

4. Are there any topics we need to clarify, or any item that was confusing?

5. Do you have any additional comments?

Please fill in your name, address, and phone number below (optional):
Name _____
Address _____
_____
Phone Number _____

Please check here if you would like us to call you ____

Thank you,

[Builder]

Appendix A

[Custom Builder]

Homeowner Manual

[Custom Builder] Homeowner Manual

Written Agreements

✔   Design Agreement–many customers rely on our expertise to guide the design of their new home and this document details that process

✔   Construction Agreement–when you are ready to build your new home, this document establishes how we will work with you to make the dream into reality

✔   Addenda–a list of the additional documents that typically accompany a construction agreement

✔   *Client Checklist*–with so many details involved, we want to confirm we remembered to provide you with all the information you will need

[Custom Builder] Homeowner Manual

Written Agreements

[Builder] works with home buyers in one of several ways to build their new homes. We offer you flexibility and the support of professionals for three approaches to home construction:

- Design-build
- Customize existing plans from our files
- Build from your plans, subject to our review and approval

Depending on the approach that is appropriate for you, we may begin our formal relationship with either a Design Agreement or a Construction Agreement. Either of these become binding only when all parties have signed all forms and attachments.

If you are new to the United States, [Builder] welcomes you and understands that you may be unfamiliar with our business procedures and traditions. We will gladly discuss any questions you may have about our written agreements and the U.S. business practices we will be following.

Design Agreement

A design agreement describes the steps we follow and the responsibilities we share with you to create your home plans. It lists the estimated costs and payment schedule for each step of this process, including such items as engineering, soil testing, and so on, as applicable.

Our design agreement covers three phases: design schematics, preliminary design, and working drawings. You meet with us one or more times during each phase to review ideas and discuss desired refinements before proceeding to the next level of detail. Between meetings you make choices, list questions, and continue to imagine your new home. More details about the design process itself can be found in Section 4 of this manual, New Home Selections.

Design Schematics

The design process begins with rough sketches called schematics drawn at 1/8-inch scale. Each 1/8-inch represents a foot, therefore a wall that will be 8 feet long when the framers build it appears as 1 inch long on these sketches. These drawings reflect our discussions about the style, size, and layout of the home you want. Informal drawings show how the proposed home fits on the lot and suggests an elevation or exterior design.

Preliminary Design

Incorporating your comments about the schematic design, the design team redraws the home, typically at ¼-inch scale. These larger drawings show more detail in both floor plan and elevations. Meetings include discussion of materials and their relative costs with you.

[Custom Builder] Homeowner Manual

Working Drawings

These plans include fully dimensioned drawings and details required for the building permit. In addition to the floor plans and elevations, working drawings include a foundation plan, electrical details, cabinet layouts, and framing layouts for floors, walls, and roof. We then obtain prices from trades and suppliers based on this information and compile a construction budget for your home plans.

After reviewing the budget, making any necessary refinements, we assemble the documents needed to apply for a construction loan and a building permit. At this point, we have fulfilled the design agreement. If you decide to proceed, we are ready to enter into a Construction Agreement.

Construction Agreement

The construction agreement is the legal document that represents your decision to have [Builder] build your new home. It describes your home (both a legal description and the street address), financing information, homeowner association information, if applicable, and additional legal provisions. We recommend that you read these documents carefully. In particular, please take note of the topics listed on our Client Checklist, which we will discuss with you prior to your signing your construction agreement.

Addenda

Several exhibits are typically attached to the purchase agreement. The features of the community where your home is to be built determine the specific items, but [Builder] uses several standard items and the list below is typical of what you will encounter.

Exhibit A

Blueprints, including engineered drawings for foundation, floor systems, or roof system; mechanical plans, framing details, cabinet layouts, and so on. Plan ownership is defined in the construction agreement.

Exhibit B

Materials and Specifications list materials and methods to be used in construction of your home.

Exhibit C

Allowance Schedule lists categories and amounts included in the price of your home for finish materials you have yet to select.

[Custom Builder] Homeowner Manual

Exhibit D

Selection Sheets outline details of your finish material choices, such as color, brand, model, and so on. Please plan to complete remaining choices within 60 days of signing your construction agreement to avoid extending the construction schedule. See Section 4, New Home Selections, for more information.

Exhibit E

[Builder] Limited Warranty, a specimen copy for your study, with the actual warranty executed at closing.

Exhibit F

Homeowner Association Documents, where applicable.

Homeowner Manual

This book is your *Homeowner Manual*. It will guide you through the building process and serve as a useful reference after you move into your new home.

Community

Our community information materials contain specific documents and disclosures about the local community.

Client Checklist

This sheet confirms that we delivered all necessary documents and discussed key topics in order to prevent surprises. Our experience shows that the new home process progresses more smoothly with good communication. To be certain that we have been clear in explaining our purchase agreement and that we have called your attention to clauses or topics that have caused confusion in the past, we will ask you to sign this confirmation at the end of the meeting.

Client Checklist

Clients _____   Date_____

Your signature below confirms that we have delivered the following items to you:

_____ Construction agreement
_____ Construction agreement addenda
_____ _____
_____ _____

_____ Approved blueprints
_____ Materials and specifications for your floor plan
_____ Allowance schedule
_____ Selection sheets for your floor plan
_____ Draw schedule (provided by lender)
_____ Limited warranty
_____ Homeowner association documents, if applicable
_____ Homeowner manual
_____ Receipts for your deposit, $_____

That we discussed the following clauses from your construction agreement:

- Allowances
- Reimbursable expenses
- Financing
- Commence and complete construction
- Change orders: procedure and schedule
- Conformance with plans and specifications
- Plan ownership
- Site visits: procedures and safety
- Noninterference
- Inspection and acceptance: orientation
- Site clean-up
- Insulation notice
- Radon disclosure
- Limited warranty: written lists for non-emergency items; standard checkpoints at 60 days and 11 months; emergency items by phone
- Homeowner association
- Settlement or closing: target delivery date and delivery date updates
- Possession
- Insurance

- ☐ Default or termination
- ☐ Alternative dispute resolution
- ☐ Co-op broker
- ☐ Entire agreement

And that we discussed the following topics to expedite communication during the process:

_____ Scheduled construction meetings
_____ Buyers' preferred contact:
     Monday – Friday _____ Phone _____
     Saturday _____ Phone _____
_____ [Builder]'s preferred contact:
     Monday through Friday, 7:00 A.M. to 6:00 P.M. at <phone>
     Saturday, 9:00 A.M. to 1 P.M. at <phone>

Other

_____
_____
_____
_____
_____
_____

Purchaser _____   Date_____

Purchaser _____   Date_____

Builder _____   Date_____

Note to Purchaser:

Store your signed design and construction agreements here.

[Custom Builder] Homeowner Manual

Budget and Financing

✔ Budget Components–properly pricing a custom home is a complex process that includes some standard elements

✔ Creating the Budget–the finished budget is used to obtain construction and permanent financing

✔ Construction Loan–temporary financing pays for materials and labor during construction, ultimately being replaced with your permanent mortgage; express or one-time close loans are usually beneficial

✔ Construction Loan Application Checklist–lists the documents and information typically needed to complete the construction loan application form

✔ Permanent Mortgage Checklist–lists the documents and information typically needed to complete the mortgage loan application form

✔ Loan Application Paperwork–an overview of the forms involved in processing your application

✔ Underwriting–key points to be aware of regarding the loan approval process, take special note of contingencies that may apply

✔ Loan Lock–lock your loan only after [Builder] has provided you with a written delivery date confirmation

✔ Loan Closing–avoid changes to your financial circumstances to protect your loan approval

✔ *Down Payment Worksheet*–to assist you in determining the amount you have available for your down payment

✔ *Monthly Payment Worksheet*--to assist you in estimating the monthly payment amount for your new home mortgage

[Custom Builder] Homeowner Manual

Budget and Financing

The more customized your home, the more complex the pricing process. Custom home clients decide hundreds of details before we can finalize the price of a home. No one can tell you what your home will cost until we've completed the design and selections. Many buyers find tolerating this ambiguity difficult. [Builder] can offer a rough budget from the beginning, but the numbers change with each choice you make. The working budget is constantly fluctuating.

Budget Components

Once you believe that most details are decided and working drawings and specifications are finalized, we begin the pricing process. Working with a list of potential trade contractors, we send each one copies of your house plans and specifications, requesting pricing information.

Reimbursable Expenses: Time and Materials

Precise bids are impossible for some categories. Trade contractors submit a "time and materials" price for work that includes unknown elements. Excavation is one example. On some sites, until the excavator begins digging, no one knows what conditions exist. The soil test reveals the composition of the soil taken from the test hole. A few feet away, conditions may be quite different.

Allowances

If you have not made final decisions in certain areas–for instance, floor coverings–an "allowance" will be included in the pricing. This gives you more time to consider these selections while a definite price for the new home can be set. If the allowance is $1500 and your final choice totals $1650, you will pay the additional $150 when you sign the order. If the total is $1400, the difference is credited to you at closing.

Change Orders

Although changes are possible during the building process, once blueprints have been drawn, engineering completed, and a building permit obtained, even a minor change can necessitate redrawing, reengineering, and reapproval by the building department and homeowner association–both time and money. Take full advantage of your design meetings to arrive at a plan that expresses your new home dream and minimize changes to avoid extra costs or extending your home's construction schedule.

Creating the Budget

We develop the house budget as prices come in. The price from each selected supplier or trade contractor is incorporated. Costs of permits, fees, taxes, insurance, and allowances are

[Custom Builder] Homeowner Manual

entered as well as a contingency amount, usually 2 to 5 percent. Real estate agent commissions, if applicable, and our margin are calculated. All these numbers lead to the (first) final budget. The (first) final budget can lead to some last-minute adjustments design or specifications, resulting in the (second) final budget.

A well-done budget is detailed, comprehensive, and realistic. This information is usually required by the construction lender as part of the loan application materials and is an essential tool for us in managing the job. The lender checks invoices against this budget in the draw process.

Construction Loan

The first loan involved in building a home is a construction loan. The construction loan serves a distinct purpose and is closely related to the permanent mortgage. The construction loan pays for materials and labor during construction.

Construction loans carry a higher interest rate than permanent financing, usually based on the prime rate plus a set percentage, commonly 1 to 2 percent. However, interest charges apply only on the amount the lender transfers to the construction account from the date it is transferred. The cost of construction interest is low in the beginning and becomes more significant as the home nears completion. Lenders usually loan up to 70 to 80 percent of the lesser of appraised value or hard costs. Hard costs include permits, materials, and labor for the work. Construction loans do not always cover the costs of builder supervision and margin.

[Builder] Financing

If [Builder] owns the lot your home is to be built on, we arrange construction financing. In that case, you arrange for permanent financing. Lenders sometimes want assurance that a permanent mortgage loan is approved to repay the construction loan when the home is complete. For this reason, the lender may require proof of approval of your permanent loan before granting the construction loan.

Streamline or Express Loans

If you own the new home site, the construction and mortgage loan will both be in your name. While you are collecting information from potential lenders, ask whether they offer a program that streamlines the application process. Sometimes called an express program, you apply and qualify for permanent and construction financing in one step. Loans can come from two different lenders, but working with one lender simplifies the process, saves time, and reduces fees.

Construction Loan Application Checklist

Whether the construction loan is obtained by you or [Builder], the items listed below are typically required by the lender. Lenders charge closing costs on the construction loan. When the construction loan closes, work on your new home can begin.

Loan Application Form

This form is provided by the lender along with a list of items needed for application processing. The loan package will also include information about the lender's draw schedule and procedures.

Credit Report

Just as with a mortgage application, the lender will check credit history.

Permanent Financing Information

The lender will want to see evidence that permanent financing is in place to pay off the construction loan when the home is completed. If you are applying for construction money and the loan is to be an express or one-time close arrangement, this is already part of the process.

Site Information

___ Legal description
___ Plot plan, survey with easements and access
___ Site status, including:
    A copy of the contract to purchase site or proof of ownership
    Subordination agreement, if applicable
    Status of taxes
    Status of title
    Zoning, appropriate approval letters
___ Soil report
___ Utilities, letter of availability
___ Permits, including as applicable:
    Well
    Septic
    Building
___ Homeowner association documents

Home Plans

___ Blueprints
    Floorplans
    Elevations
    Structural
    Mechanical
    Electrical
    Grading
    Landscaping
___ Specifications, details of finish work and materials

[Custom Builder] Homeowner Manual

Appraisal (as if the home were completed)

For the benefit of all concerned, the appraisal confirms that the completed home will be worth what you are paying for it.

Construction Budget

As bills accumulate, the lender will check each invoice against the budget category and will require an explanation of any variance. The lender will fund only items in the budget or acceptably explained. In no case will the total paid out exceed the final amount of the construction loan unless the lender reviews and approves a higher amount.

Construction Schedule

Although approximate, the lender wants to know the term of the construction loan. This affects the rate and other terms. Extensions are usually possible, but come with additional fees.

Builder Information

    ___ A copy of the construction agreement with [Builder]
    ___ Evidence of earnest money deposit
    ___ Resume on [Builder], including:
        Financial statement
        Trade and financial references
        Other work in progress
    ___ List of trade contractors and suppliers
    ___ Evidence of insurance, including:
        Builder's risk or homeowner
        Workman's compensation and general liability
        Flood insurance, if applicable

Permanent Mortgage Checklist

Your lender's job is to understand your particular financial circumstances completely. You will review all information on the application at your meeting with the loan officer. A situation rarely arises that your loan officer has not encountered in the past. Do not hesitate to discuss any questions you have regarding your assets, income, or credit. By providing complete information, you prevent delays or extra trips to deliver documents.

The amount of documentation and information required for a mortgage can seem overwhelming. You can facilitate the application process by collecting as much of the needed information as you can before your appointment.

[Custom Builder] Homeowner Manual

The checklist that follows is a general guide to assist you with the loan application. Some of the items listed may not apply to you, and your lender will probably request some items that we have not mentioned, but this list will get you off to a good start.

Credit Report

Please note that you will be asked to pay for a credit report and an appraisal upon signing the application. If you are applying for the construction loan, this will be part of that process.

Property Information

___ The purchase agreement will include the legal description of the property and the price.

Personal Information

___ Social Security number and driver's license for each borrower
___ Home addresses for the last two years
___ Divorce decree and separation agreements, if applicable
___ Trust agreement, if applicable

Income

___ Most recent pay stubs
___ Documentation on any supplemental income such as bonuses or commissions
___ Names, addresses, and phone numbers of all employers for last 2 years
___ W-2s for last 2 years
___ If you are self-employed or earn income from commissioned sales, copies of last 2 years of tax returns with all schedules and year-to-date profit and loss for current year, signed by an accountant.
___ Documentation of alimony or child support, if this income is considered for the loan

[Custom Builder] Homeowner Manual

Real Estate Owned

___ Names, addresses, phone numbers, and account numbers of all mortgage lenders for the last 7 years
___ Copies of leases and 2 years of tax returns for any rental property
___ Market value estimate

Liquid Assets

___ Complete names, addresses, phone numbers, and account numbers for all bank, credit union, 401K, and investment accounts.
___ Copies of the last three month's statements for all bank accounts
___ Copies of any notes receivable
___ Value of other assets such as auto, households goods, and collectibles
___ Cash value of life insurance policies
___ Vested interest in retirement funds or IRAs

Liabilities

___ Names, account numbers, balances, and current monthly payment amounts for all revolving charge cards
___ Names, addresses, phone numbers, and account numbers for all installment debt and approximate balances and monthly payments for such items as mortgages, home equity loans, and auto loans
___ Alimony or child support payments
___ Names, addresses, phone numbers, and account numbers of accounts recently paid off, if used to establish credit

Loan Application Paperwork

Once you have given all preliminary information to your loan officer, your lender sends verification forms to your employers, banks, and current mortgage company or landlord, and also orders the credit report and appraisal. You sign a release to authorize these steps. Your lender will provide you with a Good Faith Estimate and a Truth-in-Lending Disclosure.

Good Faith Estimate

The Good Faith Estimate lists the estimated costs you will incur at closing. Some of the numbers listed on this form are prorations, subject to change based on the actual date of the closing. Others are set fees that should remain the same.

Truth-in-Lending Disclosure

The Truth-in-Lending Disclosure shows the total cost to you, over the term of the loan, for your specific financing. The calculation is based on the assumption that you own the home and make regular payments throughout the term of the loan.

Verification of Employment

The lender sends Verification of Employment (VOE) forms to all employers for the last two years. The employers complete, sign, and return the forms to the lender. The forms show the dates of employment, the amount of money you earned last year, and how much you have earned so far this year. The VOE documents bonuses and overtime you earned.

Verification of Deposit

Verification of Deposit (VOD) forms go to each banking institution listed on your application. The institutions indicate the date you opened each account, average balances for the last three months, and the amount of money you have in each account on the day they complete the form. Any loans or overdraft accounts you have with the bank will also be shown.

Verification of Mortgage

Mortgage companies and landlords complete Verification of Mortgage (VOM) forms. These show the lender how much you owe, the amount of your monthly payment, and whether you make your payments by the due date.

Credit Report

Your credit report shows the amounts of money you owe to each of your creditors, minimum monthly payments, and your payment history. The appraisal confirms the value of the home you are purchasing for you and your lender.

Underwriting

Typically, several weeks pass as these reports and forms are returned to the lender. If any delays are encountered, the loan officer may contact you for assistance. The credit reporting agency may call you to verify that the information they have gathered is correct.

Once the loan processor has collected this standard documentation, you may be asked to write letters describing your assets, income, or credit. Few loans are finalized without requests for additional information just before the package is submitted to the underwriter for final approval. At this point you may become frustrated with the loan process.

Please remember that your lender requests these letters to assist you in obtaining your financing. Do not hesitate to discuss your concerns with your loan officer. Perhaps he or she can provide some additional insight on what may seem to be redundant requests.

Loan Amount Requested

Before the processor submits your file to the underwriters for final approval, he or she will verify the final sales price. Make sure that copies of all addenda such as change orders signed

[Custom Builder] Homeowner Manual

after the original purchase agreement was completed have been sent to the lender. This assists the lender in determining the exact loan amount. If change orders affect the total price after this point, you may have to resubmit your loan application for the higher amount or the lender may ask you to pay for the additional items in cash.

Loan Approval

During your first meeting, you and your lender determine the timing to obtain prequalification. This allows us to start the home even though final approval is still pending. You will discuss additional items that you may need to obtain final loan approval. Several weeks after your first meeting with the lender, you should receive loan approval. If any of the documents requested have not been returned to the lender in a timely manner, approval may take longer.

Contingencies

Loan approvals often carry conditions of approval. The sale of a previous home or proof of funds are two examples. Discuss any concern you may have about such conditions with your loan officer and obtain any requested documentation as soon as possible. Once all contingencies are met, the final loan can be approved.

Loan Amount Approved

If you qualify for an amount that is less than you requested, ask your loan office what changes might qualify you for a larger loan. Or, consider omitting some items now (a deck or finished basement) and adding them to your home later. Another possibility is to talk to another lender with different programs and different requirements.

Loan Declined

If, after your best efforts, your are not approved for a loan within 45 days of signing your purchase agreement, in accordance with your Purchase Agreement, [Builder] will refund your initial deposit upon you signing a release letter and returning this *Homeowner Manual* to the sales office.

Loan Lock

The only thing anyone knows for certain about interest rates is that they will change. Do not rely on anyone's predictions regarding rates. Locking your rate prematurely can result in extra expense if your new home is not complete in time to close within the lock period. We are happy to update you throughout the process of construction on the target delivery date. ***Until we reach a point in construction where factors outside our control can no longer affect the delivery date, the decision to lock your loan is at best a gamble.***

[Custom Builder] Homeowner Manual

Loan Closing

Between the time your loan is approved and the date of your closing, remember that any significant changes in your financial circumstances could impact your loan approval. If your closing occurs more than 30 days after the lender issues your loan approval, the lender may order an additional credit report just prior to the closing date. Changes in your financial circumstances, for example, purchasing a new car or increases in your charge card will appear as a new liability on your updated credit report. Such changes may cause your lender to reconsider your approval. Holding off on such purchases until after closing is best.

[Custom Builder] Homeowner Manual

Down Payment Worksheet

Available Funds

Equity in present home	$	_____
Savings, savings certificates		_____
Investments		_____
Insurance (cash value)		_____
Other funds (such as a cash gift)		_____

Total available funds _____

Minus amount you want to keep in savings _____

Adjusted Total Available Funds $ _____

Expected Expenses

Settlement costs (estimate 5 percent of loan)	$	_____
Moving costs		_____
Landscaping		_____
Other expected expenses		_____

Total Expected Expenses $ _____

Down Payment

Adjusted total available funds $ _____
Minus total expected expenses _____

Amount Available for Down Payment $ _____

[Custom Builder] Homeowner Manual

Monthly Payment Worksheet

Loan Payment

    Principle and interest $ _____
    Property tax   _____
    Hazard insurance   _____

    Total Loan Payment $ _____

Homeowners Association Monthly Dues $ _____

Estimated Utilities

    Electric $ _____
    Gas   _____
    Water   _____
    Sewer   _____
    Trash collection   _____
    Cable TV   _____
    Security system monitoring   _____

    Total Estimated Utilities $ _____

Monthly Payment

    Loan payment $ _____
    Homeowner association dues   _____
    Estimated utilities   _____

Total Monthly Payment $ _____

[Custom Builder] Homeowner Manual

New Home Design and Selections

✔ Selecting a Homesite–[Builder] may be helpful in locating an appropriate homesite or suggesting how best to work with one you have already selected

✔ Design Considerations–the past, present, and future should come together in your new home, combining lifestyle and design to create a comfortable environment for your family

✔ *Homesite Checklist*–a guide to avoid forgetting important considerations

✔ *Design Details*–this list will start you thinking about the details for your new home

✔ Selection Hints–reminders to guide you through the selection process

✔ Selection Locations–names and locations of showrooms where you can view selections and options

✔ Change Orders–[Builder] will consider requests for changes after you sign the Buyer Start Order in accordance with the schedule and fees described here

✔ *Change Order*–a copy of the form that documents any changes, requiring the signatures of all parties and full payment prior to being implemented

[Custom Builder] Homeowner Manual

New Home Design and Selections

Building a custom home includes finding an appropriate location, designing the home, and selecting features, finish materials, and colors. As you make these decisions about your new home, consider your present and future lifestyle. Take into account your family's daily activities, hobbies, and work; the kind of entertaining you do, and your family's holiday traditions.

Selecting a Homesite

If you are looking for a homesite, [Builder] may be able to assist you. We may have a lot in inventory or have access to a building site that would suit your needs. As a licensed real estate broker, [Builder] can negotiate your land purchase as well. Perhaps most importantly, we can analyze the site to identify special features or concerns that could impact the design of your home or the budget.

Compare the possibilities for the site to your lifestyle. If your family enjoys morning coffee on the deck, outdoor activities such as gardening, children's games, or sunbathing, will this site adapt to those activities? Will the house you want to build fit on the lot? Is the lot appropriate for the style you have in mind? If you have not completed your house plans, information about the lot may affect design decisions. Consider shape, size, angles, and massing of materials to blend your home design with a particular site. [Builder] can help you analyze the site and make all these determinations. The Homesite Checklist, which can be found in this section, can also help guide your final choice.

Design Considerations

Just as in selecting a location, as you design your home and makes selections, analyze the needs of your family, as a group and as individuals. Do you prefer formal or informal entertaining? Do you host out-of-town guests? Think about now and think about the future. What changes do you expect in the next 5 to 10 years? Identify the details that change a house into a home for your family. The right house for your family is one that rekindles the best memories as it provides a setting for new ones.

Homeowner Association Design Review Committee

[Builder] builds many homes in covenant-protected communities where the homeowner association design-review committee uses criteria established by the association to review home plans. The goal is to assure that homes in the community meet agreed-upon standards affecting size, design, exterior finish materials and colors, and height. If your new home will be in a covenant-protected community, be certain that everyone on your design team is informed about the criteria your association uses in its design-review procedures.

Environmentally Friendly Construction

Materials and methods in new home construction are changing and improving every day. Efforts to use resources efficiently have led to the development of many "alternative materials" which take the place of wood in the home. Alternative materials offer choices that affect cost, the construction schedule, and long-term performance as well as the home's impact on the environment. [Builder] will be happy to discuss the pros and cons of all such choices with you.

Energy efficiency can be maximized when planned into the home from the beginning. Passive and active solar can save energy. And while the federal government has mandated water-saving devices must be installed in all new homes, xeriscape landscaping remains a choice for the homeowner–one that can save work and dollars.

[Builder] has implemented many practices that reduce waste during construction, and we recycle where possible. Similarly, we encourage our clients to plan for convenient recycling in their home plans.

Modern Technology

With the advent of the Internet and increased popularity of the personal computer, home technology has also likewise become increasingly important. Design into your home the capability for present and future communication, entertainment, and efficient operation. Smart House features may be for you. Security, lighting, temperature control, ventilation, and many maintenance activities can readily be automated for convenient home management.

Document Your Ideas

Many clients find it helpful to assemble a scrapbook of sketches, photos, model numbers, samples, and color chips. Gather notes and ideas for each room or area of the home. A partial list of the Design Details you will make choices about can be found in this section, to get you started.

Set Priorities

Each decision for your new home influences cost and impacts other choices as well. How important is that three-car, side-entry garage? A side-entry garage will require a different driveway shape than a front-entry garage. This will affect the landscaping plan. Or would you rather apply the cost difference to ceramic tile countertops and upgraded carpet? Think of this process not as giving up things you want, but of determining which items are essential and which are preferences. If something must be omitted, select an item you can add later, such as wallpaper, rather than sacrificing something that is impractical to add later, such as a curved staircase. [Builder] can help with information on what is practical to add later and what is not.

Homesite Checklist

Cost
- ☐ Lot price
- ☐ Property taxes

Legal Status
- ☐ Zoning (subject to change)
- ☐ Building department
- ☐ Homeowner Association
- ☐ Owner
- ☐ Owner's agent
- ☐ Title

Survey
- ☐ Size
- ☐ Boundaries, property corners
- ☐ Easements
- ☐ Setbacks
- ☐ Square footage requirements
- ☐ Height restrictions
- ☐ Orientation

Fees
- ☐ Permit
- ☐ Water tap fee or well cost
- ☐ Sewer tap fee or septic cost
- ☐ Electric hook-up
- ☐ Homeowner association deposits or fees
- ☐ Gas hook-up
- ☐ Phone hook-up
- ☐ Cable TV or satellite service

Unique Construction Factors
- ☐ Access
- ☐ Foundation required for soil conditions
- ☐ Site preparation (trees, rocks, groundwater)
- ☐ Special grading concerns
- ☐ Seasonal weather impact on schedule
- ☐ Ecological concerns
- ☐ Adjacent sites/view

Hazards or Concerns
- ☐ Flood plain
- ☐ Weather
- ☐ Seismic zones
- ☐ Crime rate
- ☐ Pollution

Services
- ☐ Government
- ☐ Transportation
- ☐ Schools
- ☐ Cultural Amenities
- ☐ Post office and mail delivery
- ☐ Police and fire protection
- ☐ Road maintenance, snow removal
- ☐ Trash collection
- ☐ Banking, Business opportunities
- ☐ Places of worship
- ☐ Recreation

[Custom Builder] Homeowner Manual

Design Details

Listed here are some typical issues we need to made choices about for your new home. Begin to review the list and, as you do, keep a list of questions that come to mind.

Foundation
- Poured concrete
- Insulation on exterior
- Insulated forms on interior and exterior

Exterior Wall Coverings
- Standard stick framing
- Structural insulated panels

Interior Wall Framing
- Stick framing
- Metal stud framing

Roof Framing
- Trusses
- Stick framing

Floor Framing
- Wood "I" joists
- Floor trusses
- Standard dimensional lumber

Exterior Wall Coverings
- Siding
  __Masonite, other manmade material
  __Cedar or redwood
  __Fiber cement products
  __Plywood
  __Cedar shingles
- Masonry
  __Brick
  __Natural stone
  __Man made stone
- Stucco

Roof Coverings
- Composition shingles
- Cedar shakes
- Metal shakes
- Fiber cement shakes
- Tile
- Metal
- Slate

Windows
- Wood with exterior cladding
- Wood with fiberglass exterior
- Vinyl

Exterior Doors
- Entry
  __Wood
  __Steel
  __Fiberglass
- Garage side door
- Patio door
- Other doors

Overhead Garage Doors
- Size (standard, oversize)
- Material (metal, wood, vinyl)
- Insulated?
- Glass in door(s)?
- Garage door opener(s)

Decks
- Redwood
- Cedar
- Composite
- Molded plastic products

Plumbing: Review catalogs or showrooms for sinks, toilets, faucets, and tubs. Note the brand name, model or part number, color, and so on.

Electrical
- Phones
- Cable TV
- Computer modems
- Media center
- Ceiling fans
- Dimmer switches
- Special circuits (garage/hobby/office)
- Exterior outlets

- Indirect lighting
- Undercounter lighting
- Intercom system
- Security system
- Central vacuum system

Heat, Ventilating, and Air
- Gas forced air
  __80% efficient
  __90%+ efficient
- In-floor radiant heat
- Baseboard hot water heat
- Heat pump
- Air conditioning
- Swamp cooler
- Humidifier
- Electronic air cleaner
- Day-nite thermostat

Fireplace(s)
- Wood burning (metal, mason)
  __High-efficiency metal fireplace (EPA phase II)
  __Recirculating fan
- Gas burning
  __Recirculating fan
  __Control (manual, switch, remote)
- Fireplace surround
  __Brick
  __Stone, real or man made
  __Tile
- Mantle, design and finish

Insulation
- Exterior walls
  __Foam core panels
  __Fiberglass
  __Cellulose
  __Spray in place foam
- Walls or ceilings (for sound)
- Garage insulation
  __Walls
  __Ceilings
  __Overhead doors
- Unfinished basement walls

209

[Custom Builder] Homeowner Manual

Drywall
- ☐ Texture style
  - __Knockdown
  - __Orange peel (splatter)
  - __Stomp brush
  - __Southwest
  - __Smooth finish
- ☐ Corners (square, round)
- ☐ Other drywall treatments
- ☐ Garage

Interior Trim
- ☐ Door style
- ☐ Door/window casing
- ☐ Window treatment
  - __Drywall jambs with sill
  - __Wood jambs with casing
- ☐ Baseboard
- ☐ Built ins
  - __In closets
  - __Shelf units, bookcases
  - __Media centers
  - __Desk
- ☐ Crown molding
- ☐ Chair rail
- ☐ Window seats
- ☐ Closet arrangements
  - __Rods (single, double)
  - __Shelves (fixed, adjust)
  - __Cedar lining
  - __Shoe rack(s) or shelves
  - __Wire closet arrangements
  - __Pre-built melamine
- ☐ Handrails
  - __Style/material
  - __Finish

Cabinets
- ☐ Brands/style/color
- ☐ Extras
  - __Roll out shelves
  - __Glass door inserts
  - __End panels match doors
  - __Hardware
  - __Spice rack
  - __Wine rack
  - __Book shelves
  - __Cabinets as furniture
  (hutch, bookcase, etc.)

Painting
- ☐ Exterior
  - __Body color
  - __Trim color
  - __Accent color
  - __All stain
  - __Sheen (flat, stain)
- ☐ Interior trim
  - __All paint
  - __All stain
  - __Combination
- ☐ Ceilings
- ☐ Walls
- ☐ Garage, if drywalled

Countertops/Backsplashes
- ☐ Kitchen
  - __Island
  - __Kitchen other
- ☐ Wet bar
- ☐ Powder room
- ☐ Master bath
- ☐ Bath #1
- ☐ Bath #2
- ☐ Bath #3
- ☐ Laundry

Mirrors, Shower Doors
- ☐ Mirrors
- ☐ Shower/tub enclosure (clear, frosted, special, brass, chrome frame)
- ☐ Medicine cabinets
  - __Powder room
  - __Master bath
  - __Bath #1
  - __Bath #2
  - __Bath #3
- ☐ Mirrored closet doors

Appliances
- ☐ Gas or electric?
- ☐ Stove unit with oven
- ☐ Cooktop, separate oven(s)
- ☐ Oven (single, double)
  - __Convection
  - __Self/continuous clean
- ☐ Microwave oven
- ☐ Refrigerator
- ☐ Freezer
- ☐ Trash compactor
- ☐ Dishwasher
- ☐ Washer
- ☐ Dryer

Wall tile or wall surfaces
- ☐ Master bath
- ☐ Bath #1
- ☐ Bath #2
- ☐ Bath #3

Floor Coverings
- ☐ Entry
- ☐ Living room
- ☐ Dining room
- ☐ Kitchen
- ☐ Nook
- ☐ Hallway
- ☐ Great room
- ☐ Office or study
- ☐ Master bedroom
- ☐ Master bath
- ☐ Bedroom #2
- ☐ Bedroom #3
- ☐ Bedroom #4
- ☐ Storage room
- ☐ Stairs

Hardware
- ☐ Entry handle set
- ☐ Exterior lock sets
- ☐ Interior knobs
- ☐ Door stops
- ☐ Towel bars/towel rings

Miscellaneous
- ☐ Retaining wall material
- ☐ Window wells
- ☐ Downspout/Splash Blocks

Selection Hints

[Builder] provides you with selection sheets that list the choices you need to make. Schedule time to visit both our office and our suppliers' showrooms to make your selections as soon as possible. Plan to finalize your selections within 60 days of signing your construction agreement. Your prompt completion of these selections helps prevent delays caused by backorders.

Informed Choices

We recommend that you review the maintenance tasks and warranty guidelines in Section 8 of this manual prior to making your selection decisions.

Be Thorough

Our selection sheets are very detailed. Fill in all blanks completely. Costly errors arise from assumptions and incomplete selection sheets. After completing this form, double-check all color numbers and names and sign and date each page.

Allowances

Decorating choices that exceed the specified allowances, such as those for floor coverings, countertops, or light fixtures, will require additional payment. Although such amounts can be credited to you at closing and subsequently added to your mortgage, they are not refundable.

Colors

You are welcome to bring cushions or swatches to showrooms to coordinate colors. View color samples in both natural and artificial light to get an accurate impression of the color. Variations between samples and actual material installed can occur. This is due to the manufacturer's coloring process (dye lots) and to the fact that over time, sunlight and other environmental factors affect the samples. Some colors appear different when seen in a large area as opposed to the sample.

Exterior Choices

Your homeowner association and selections your future neighbors have made may limit your choices for exterior finish materials or colors. The sooner you can make your selections, the more choices you have. Viewing existing homes is one way to select exterior colors. Selections often look different on a full-size home. Some colors require extra coverage which can impact the cost.

Selection Hold

We reserve the right to place a hold on your selections until your lender has approved your loan and all contingencies are released. If suppliers have discontinued any of your selections, we will contact you and ask you to make an alternate selection within 5 days. Occasionally, a home is already under construction and [Builder] has made some or all of these choices.

Availability

If a selection you make turns out to be unavailable, we will contact you and request that you make a different selection within five business days. Because so many choices are offered, [Builder] is unable to predict when a particular manufacturer or supplier may discontinue any particular item. We regret any inconvenience this causes. Similarly, materials readily available when your home is built may not be available in years to come if replacements are needed.

Record of Selections

Please retain your selection sheets for future reference. They are useful for matching paint colors, tile grout, and replacement items in your home.

Selection Locations

Item(s) _____
Contact _____
Company _____
Address _____

Phone _____
Fax _____
Hours _____

Item(s) _____
Contact _____
Company _____
Address _____

Phone _____
Fax _____
Hours _____

Item(s) _____
Contact _____
Company _____
Address _____

Phone _____
Fax _____
Hours _____

Item(s) _____
Contact _____
Company _____
Address _____

Phone _____
Fax _____
Hours _____

Change Orders

[Builder] uses a change order form (see sample at the end of this section) to describe and document all changes you may request to your new home's plans and specifications. Change orders fall into three categories. You may decide to:

- Add or delete an item after signing your selection sheets
- Change a selection previously ordered
- Personalize your home plans still further with a custom-designed feature

In order to deliver your home as close as possible to the target date, we order many items well in advance of installation. Once a particular item is ordered, making further changes may involve an adjustment in the planned delivery date and additional costs. We will itemize these for you as part of the change order documentation process.

Processing

When you request a change, [Builder] documents the request and submits it for approval and pricing. Pricing of change requests typically takes 5 to 10 business days.

Design/Pricing Deposit

Your first three change requests will be designed and priced without a design/pricing deposit. Beginning with the fourth change request, a design/pricing deposit of $300 will apply. The full amount becomes a credit against the cost of the change if you approve the change order. If you decide not to proceed, [Builder] retains the design/pricing deposit.

Sometimes a seemingly minor change impacts other elements of the home and therefore may come with hidden costs. For example, if you order a ceiling fan, the framing that will hold it is reinforced. If you add a window, framing, drywall, interior and exterior trim, and paint costs may all be affected.

Administrative Fee After Cutoffs

Changes after the cutoff dates set for your home's construction include an administrative fee. This is necessary because previously issued paper work must be canceled and reissued. Errors in this process are a [Builder] responsibility. If the change you request impacts the schedule, construction our pricing will include construction loan interest for the additional days. The cost of deleted items will be credited to you although administrative fees are non-refundable.

[Custom Builder] Homeowner Manual

Information on pricing and any schedule adjustment is returned to you for a final decision. If you elect to proceed with the change, we ask that you sign the change order and make full payment. Change orders that remain unsigned or unpaid become null and void upon the expiration date shown on the change order.

For the protection of all concerned, all changes are documented and incorporated into your new home only after:

- [Builder] has approved and signed the change
- You have approved, signed, and paid for the change prior to its expiration date
- The applicable building department has approved the change

Our contracts with our trade contractors prohibit them from making any changes to plans or specifications without written change order authorization from [Builder].

Cutoff Points for Changes

[Builder] follows a schedule of cutoffs for changes as shown below. [Builder] reserves the right to deny changes you request after these cutoffs.

Changes affecting	Should be made prior to the start of
Foundation	Engineering and permit application
Windows, doors, elevation	Foundation
Mechanical systems, cabinets, appliances	Framing
Texture, hardware, lighting	Mechanical rough-ins
Interior trim and floor coverings	Insulation
Landscape design or materials	Interior trim

[Builder]

Change Order # _____

Purchasers _____ Date _____
Contract dated _____ Plan _____
Address _____ Lot # _____

ature *Description of Change*

_____
_____
_____
_____
_____
_____

Design/pricing deposit _____   Expiration date _____
Administrative fee   _____
Cost of change   _____   Delivery date adjustment _____ days
Credit (deleted items) _____
Total   _____

Purchasers have requested the change described above, its costs, and the corresponding adjustment in the construction schedule. By signing this change order, Purchasers agree to pay for this change and acknowledge that the estimated delivery date for the home is revised accordingly. [Builder] will incorporate the change into the home only when the change order has been approved and signed by [Builder], and paid in full by Purchasers. [Builder] has the option of revising the cost, delivery date adjustment, or both if Purchasers have not signed this change order by the expiration date above.

Approved,_____   Purchaser_____
             [Builder]
                                                           Purchaser_____

Date_____   Date_____

Date payment received_____

Note to Home Buyer:

Insert your records of your new home selections here.

[Custom Builder] Homeowner Manual

Construction of Your Home

✔ Preconstruction Conference–a meeting to review your plans, selections, changes, and the protocols of the construction process

✔ Start of Construction–as anxious as we all are to get started, [Builder] must attend to several tasks before construction begins

✔ Safety–please respect the potentially dangerous nature of a construction site and follow our site visit policies

✔ Frame Tour–your second meeting with your builder provides an opportunity to see the quality inside the walls of your new home and confirm that selections and change orders are correct so far

✔ By Appointment Tours–you are welcome to set appointments for an on-site visit where you will have our undivided attention to discuss any questions

✔ Locks and Keys–once you use your house keys, only your keys will open your home

✔ Plans and Specifications–no two homes are alike

✔ Quality-- we monitor work on your home to note and correct any errors that occur and ensure that the home we deliver meets the standards we promised you

✔ Single Source–[Builder] selects all personnel and orders all materials that go into your home

✔ Trade Contractors–trades people have no authority to make changes without [Builder] written change order and are unaware of all the elements in your home; any questions you have should be communicated through your sales person

✔ Schedule–delivery dates are a target until we confirm a closing date in writing; we promise a minimum of 30 days notice

✔ Construction Sequence–an overview of the major steps typically followed in building a home

✔ *Our Customer Wants to Know*–forms for your convenience, please document any questions you have about your home during construction and forward them to your sales counselor

[Custom Builder] Homeowner Manual

Construction of Your Home

The construction of a new home differs from other manufacturing processes in several ways. By keeping these differences in mind, you can enjoy observing the construction process as we build your new home.

- As a consumer, you rarely have the opportunity to watch as the products you purchase are created. Your new home is created in front of you.
- You have more opportunity for input into the design and finish details of a new home than for most other products. Our success in personalizing your home depends on effective and timely communication of your choices.
- Because of the time required for construction, you have many opportunities to view your home as it is built, ask questions, and discuss details.

Preconstruction Conference

Just prior to the start of construction we will ask you to meet with us for a comprehensive review of your final plans and specifications as well as the building process itself. We discuss site visits, questions, trade contractor communication, change orders, and target delivery date. A copy of our agenda is included on the next page. Please bring any questions and this manual.

Start of Construction

Construction will begin several weeks after this meeting. Before construction of your home can begin, [Builder] has several important tasks to accomplish that involve outside people and entities. For example, in order to apply for a building permit, we assemble documents and approvals from many sources:

- ✔ Permit application form
- ✔ Site plan showing the house on the lot, the easements, and setbacks
- ✔ Complete set of plans
- ✔ Soil report
- ✔ Engineer-stamped set of foundation plans
- ✔ Engineered truss drawings
- ✔ Letter from water and sewer district
- ✔ For a septic system, percolation test results
- ✔ For a well, a well permit
- ✔ Rural lots may require a driveway permit to establish access to a county road
- ✔ A letter of acceptance from the homeowner association, if applicable

Preconstruction Conference Agenda

Client _____ Date _____

Address _____

At the office:

1. Site plan
2. Soil report
3. Drainage plan
4. Status of permit
5. Utilities status
6. Status of Design Review approval
7. Other HOA issues
8. Submit landscape plans to Design Review
9. Blueprints
   a. Elevations
   b. Floor Plan
   c. Cabinet layout
   d. Mantel
   e. Electrical
   f. Other
10. Specifications
11. Selections and options as of this date
12. Timetable for remaining selections
13. Change orders
14. Change order cutoff schedule
15. Target start date
16. General construction sequence/schedule
17. Events that will extend schedule
18. "Nothing's happening at the house"
19. Quality, builder's inspection of work
20. Site visit guidelines
    a. Safety
    b. Security
    c. Work in progress
21. How to handle questions
22. Scheduled site visits
23. Draw schedule and procedures
24. Payment due at this time
25. Next payment due date
26. Target delivery date
27. Reminder: read maintenance and warranty information
28. Other [Builder] topics:

29. Other Client topics:

30. Tentative time for next meeting

At the site:

31. Lot boundaries
32. Set backs
33. Easements
34. Orientation of home
35. Trees and other natural features
36. Drainage issues
37. Drive / culvert
38. Utility trenches
39. Construction
    a. Utilities
    b. Trash
    c. Sanitary facility
40. Mail box

Client _____ Client _____ Builder _____

[Custom Builder] Homeowner Manual

Safety

We understand that you will want to visit your new home between these construction reviews. A new home construction site is exciting and can also be dangerous. Your safety is of prime importance to us. Therefore, we must require that you contact [Builder] before visiting your site. We reserve the right to require that you wear a hard hat and that a member of our staff accompany your during your visit. Please observe common-sense safety procedures at all times when visiting:

- Keep older children within view and younger children within reach, or make arrangements to leave them elsewhere when visiting the site.
- Do not walk backward, even one step. Look in the direction you are moving at all times.
- Watch for boards, cords, tools, nails, or construction materials that might cause tripping, puncture wounds, or other injury.
- Do not enter any level of a home that is not equipped with stairs and rails.
- Stay a minimum of 6 feet from all excavations.
- Give large, noisy grading equipment or delivery vehicles plenty of room. Assume that the driver can neither see nor hear you.

In addition to safety considerations, be aware of the possibility that mud, paint, drywall compound, and other construction materials are be in use and can get onto your clothing.

Frame Tour

Many buyers appreciate the opportunity to tour their home just after the rough mechanical stage, before insulation. The rooms have begun to take shape but the inner workings are still visible. This is an opportunity for you to see the quality that goes inside the walls of your home.

This meeting also gives us an opportunity to confirm that we are correctly installing the items you ordered or approved changes you requested. We will also update you on the target delivery date during the frame tour. Frame tours usually take 30 to 45 minutes. Please remember to bring this homeowner manual, selection sheets, and any change orders. If for any reason you are unavailable to attend this meeting, we must continue with construction.

By-Appointment Tours

In addition to the two routine meetings on site, you are welcome to schedule time with your builder to view your home together and discuss any questions you may have. Please call to set an appointment so your builder can arrange his or her schedule to give you undivided attention.

Appendix B

Warranty Service:

Optional Procedures

Warranty Service: Optional Procedure

An alternative to the traditional "send in a list" approach to warranty can have a positive and powerful impact on your customers' opinion. Consider substituting the segment that follows for the "60 day Report" that appears in the standard homeowner manual model.

60-Day Warranty Visit

During your homeowner orientation, [Builder] sets a tentative appointment with you to revisit your home in approximately 60 days. This follow-up visit has several purposes. During this meeting we will:

- Review key maintenance points and answer any questions you have about the care and operation of your home's features.
- Inspect your home, using our checklist as a guide, to confirm all of the components are performing as we intend them to.
- Inspect any items you believe require warranty attention to determine appropriate action.

Warranty Service Request Forms. To make keeping a list of items more convenient and to ensure that we have the necessary details to respond promptly, we have included warranty service request forms at the back of this section.

Confirmation Call. [Builder]'s warranty office will contact you 5 to 10 days prior to the appointment to confirm the date and time. At that time we will also ask that you forward your list, if you have noted any items, so that we can do any necessary research and appropriate amount of time on our schedule. If you happen to note any additional items between then and your appointment, we will add it to you list as we go through your home with you.

[Builder Logo]

Warranty Check-up

Name _____   Date _____
Address _____   Community _____
Ph-Home _____   Lot # _____
Ph-Work _____   Plan _____
Ph-Work _____   Closing date _____

___60-Day   ___11 month   Inspection Date _____

....

- ☐ Backfill
- ☐ Drainage
- ☐ Downspout extensions
- ☐ Concrete flatwork
- ☐ Front Door
    Lock and deadbolt
    Threshold
    Weatherstrip
    Doorbell
- ☐ Back Door
    Lock
    Threshold
    Weatherstrip
    Patio door lock
- ☐ Garage overhead door
- ☐ Smoke detectors
- ☐ Furnace filter
- ☐ Interior doors
- ☐ Interior trim
- ☐ Cabinets
- ☐ Tile
- ☐ Caulk
- ☐ Window operation
- ☐ Drywall
- ☐ Floor coverings
- ☐ Homeowner list?

By _____